DIGITAL PROJECT MANAGEMENT

The Complete Step-by-Step Guide to a Successful Launch

TAYLOR OLSON, PMP

Copyright © 2016 by Taylor Olson

ISBN-13: 978-1-60427-125-6

Printed and bound in the U.S.A. Printed on acid-free paper.

10 9 8 7 6 5 4 3 2 1

Library of Congress Cataloging-in-Publication Data

Olson, Taylor, 1969-
  Digital project management : the complete step-by-step guide to a successful launch / by Taylor Olson.
    pages cm
Includes index.
  ISBN 978-1-60427-125-6 (hardcover : alk. paper)
  1. Project management. 2. Information technology projects—Management. I. Title.
  T56.8.O47 2015
  004.068'4—dc23   2015022778

Direct all inquiries to J. Ross Publishing, Inc., 300 S. Pine Island Rd., Suite 305, Plantation, FL 33324.

Phone: (954) 727-9333
Fax: (561) 892-0700
Web: www.jrosspub.com

DEDICATION

This book is dedicated to my parents, Jim and Char, who have always loved and supported me.

CONTENTS

ACKNOWLEDGMENTS

I'd like to offer my many thanks and sincere gratitude to the following people who helped me put this book together. Each of you responded to my requests for assistance with excitement and enthusiasm, and I really appreciate your professional guidance. I'm truly blessed to count you as my friends and colleagues.

Jeff Gilliard

Elaine Griffin

Joey Groh

Jennifer Harrison

Scott Hensler

Jerry Howcroft

Melissa Knisely

Kendra Maluchnik

Daniel Murphy

Paul Olson

Grace Tsai

Special thanks to Drew Gierman at J. Ross Publishing, who took a rookie and patiently taught her the ropes. Thanks for believing in me more than I believed in myself.

FOREWORD

With the digital landscape growing, changing, and evolving by the minute, the ability and need to have a concise understanding of *digital* delivery grows exponentially. Since the early 2000s, when organizations were driven by large screen desktop deployments—to today, where *mobile first* is the mantra, every project manager has had to change, quickly, to make our collective digital ecosystem thrive.

Digital Project Management: The Complete Step-by-Step Guide to a Successful Launch presents the clearest understanding of the delivery process—not in theoretical terms, but in real-world, practical terms that will enhance the knowledge and skills of anyone associated with the rollouts of digital projects.

From driving an effective gap analysis, to the details of a proper business requirements document, to budgets, and to a rich education on the technical elements required, the 30 essential steps laid out in this valuable guide provide more than a simple step-by-step procedure, they deliver real depth and education as to *why* these are important—gained from real-life experience and learning.

Having had the good fortune of working with Ms. Olson, I have been lucky to witness her expertise in action—leading not only experienced players in the business, but nurturing newcomers as well. When she asked me to read the manuscript of her book on digital project management and share my thoughts with all of you, I could not have been more honored.

Taylor Olson has been at the center of this digital evolution, and has driven not only procedural steps, but real insight into the value, requirements, and role of the project manager in the delivery process. Living at the center of the maelstrom, her thinking has proven to be invaluable to all levels and disciplines involved in the delivery of both complex and simple rollouts.

Digital Project Management: The Complete Step-by-Step Guide to a Successful Launch is an enjoyable read that shares a lot of valuable experience-based knowledge. I encourage each of you to dive in and learn how it is done from a world-class professional. Enjoy!

Bill Kolb
Chairman and CEO, Commonwealth//McCann
Chairman, MRM//McCann

ABOUT THE AUTHOR

Photograph by Kendra Maluchnik

Taylor Olson, PMP, has more than twenty years of experience in project management. She started her career in the demanding field of automotive marketing and advertising for both small and large agencies, including BBDO Detroit and McCann Worldgroup. She has traveled the world to train and collaborate with global team members in South America, China, Germany, Thailand, Canada, Mexico, Bermuda, and just about every major market in the United States, documenting best practices along the way.

Taylor graduated from Michigan State University in 1992, just as the Internet was beginning to be defined and governed throughout the world. Her automotive clients were keen on early adoption and thus, she became part of the historic transition from print to digital. Taylor currently specializes in using content management systems to produce and maintain global web solutions for large corporations operating in multiple countries and languages.

Throughout her career, Taylor has been focused on mentoring other project managers and giving back to the community through both educational and faith-based programs. She is honored to be part of Leadership Oakland Class LOXV, and was a member of the Board of Directors for the Rochester Community Schools Foundation.

Find Taylor on LinkedIn or on Twitter @RolloutManager.

At J. Ross Publishing we are committed to providing today's professional with practical, hands-on tools that enhance the learning experience and give readers an opportunity to apply what they have learned. That is why we offer free ancillary materials available for download on this book and all participating Web Added Value™ publications. These online resources may include interactive versions of material that appears in the book or supplemental templates, worksheets, models, plans, case studies, proposals, spreadsheets and assessment tools, among other things. Whenever you see the WAV™ symbol in any of our publications, it means bonus materials accompany the book and are available from the Web Added Value Download Resource Center at www.jrosspub.com.

Downloads for *Digital Project Management: The Complete Step-by-Step Guide to a Successful Launch* consist of:

- Printable PDF files for the engagement process, change control process, and the digital rollout process checklist
- Editable Word files for the business requirements document and two versions of status reports
- The digital project schedule followed throughout the book provided in both Microsoft Project and Excel
- Cutover management tools including a 301 redirects example file and complete cutover plan
- All of the other documentation and tools described in the book, including editable templates for budget estimates, change control log, content tracker, issues log, risk register, stakeholder list, and user test scripts
- A Leader's Guide is available for free for adopting instructors. This may also be purchased for non-academic use.

INTRODUCTION

Thinking back on your career as a project manager (PM) at a digital agency, how many times have you been provided with a process by which to do the assignments you've been given? I doubt this happens very often for ongoing projects, let alone big initiatives that are new to the agency. So many times we're given our assignments with little to no direction or process documentation, and we need to figure it all out for ourselves as the project moves along. It's like there's a train speeding down the line and we're trying to lay the rest of the track down before it catches up to us.

Over the past several years I've been helping to build global websites and applications for Fortune 50 corporations. It's been the best experience of my career, and now that I've learned how to do it correctly, I want to share that knowledge with others. This book lists and explains what I have found to be the *30 essential steps to ensure a successful digital project*. This process, what I call the *digital rollout process*, may be applied to a number of different digital scenarios, but to help explain the steps and concepts, I've created a fictional company called Jetzen Spaceships, who have asked our fictional agency to build a responsive, customer-facing web solution using mobile first methodology.

Before you cringe and decide that 30 steps are just too many to take back to the team at the agency, let me tell you that I know it sounds like a lot, but it is what it is. This is a how-to book for the *on-the-ground* team—not a book about theory or marketing strategy. There are 30 steps, and I've personally used this process on about one hundred digital projects. It works. I can't guarantee a successful launch for any project because there are just too many things outside of the project manager's control that can go awry—but this process will help the project manager plan every aspect of web development before launch occurs. Figure I.1 shows just a few things the project manager needs to consider and coordinate along the line.

This book has been written with every experience level in mind—from beginner level to oldsters like me, who have been doing this for

Figure I.1   Digital project managers need to know a little bit about a lot of different topics

years. Incorporating the primary project management trifecta of schedule, budget, and resourcing with newer methodologies like *responsive design* and *mobile first* guarantees this book applies to everybody working in the field. While beginners may want to follow the steps one-by-one as they move through their own projects, experienced PMs can get some new ideas and check their current practices against the ones I list in this book. I've been working with some of the largest companies in the world, and these companies have substantial digital budgets—so this process has had to measure up to huge expectations.

Reading this book will help project managers understand:

- What it takes to put together a thorough scope document, starting with a gap analysis and including budget and deliverables
- Task-level details of a typical project plan
- How information architects work with the creative team to produce a responsive content strategy
- When mobile first methodology is applied within the process and which deliverables are affected
- What a project manager needs to know about search engine optimization, analytics, infrastructure management, and testing
- Every aspect of cutover management including prelaunch steps and an example of a launch-day activities schedule
- What the potential risks are for this type of project and how to avoid or control them
- How to apply an agile approach to design and development in a setting that demands clear requirements and deliverables

The digital world is growing and changing at a rate that can seem overwhelming to those of us who have to keep up with it to build customer-facing solutions and applications. Just when we figured out how to create desktop solutions that also provided a responsive experience for mobile users, now we're flipping the entire process upside down by starting with the mobile design (mobile first) and enhancing that experience for desktop or tablet users. At this point, it's hard to predict what will be coming down the track next, but I do know that *something's* coming—there will always be something blazing down the railway.

The great thing about this process is that no matter how the technology changes, the same process can be applied with just minor adjustments. For instance, this book explains what wireframes are and why they're important. The way we create wireframes might change, and the projects we're applying them to may change, but we'll always need to plan out the framework of a design before moving forward.

I originally developed the rollout process to provide myself and my colleagues around the globe with a tool we could use to ensure we were all following a standardized process, and to make sure that we didn't forget anything along the way. I've been lucky to work with and travel to many global locations and have been involved in several workshops with my global team members to learn how they approached the work. The 30-step process in this book incorporates each of the best practices I've learned through research and my own work experience. PMO officers are also encouraged to learn and embrace the 30 steps, and to merge them with their current standards and methodology. I've even included tons of examples of documentation, along with instructions on how to create them and why they're important. My colleague, Jennifer Harrison, explains how learning the process has helped her:

> *"Using the rollout process for complex technical projects enables me to break project work into logical and manageable steps from kick-off to launch. Following along with the process ensures that I don't miss critical steps that might otherwise be overlooked, and provides a guideline for managing complicated work across functional teams. Projects small and large, local or global, and with varying scope can all greatly benefit from the structure provided by this rollout process."*
>
> —Jennifer Harrison, Digital Project Manager

Read along as our agency receives a phone call from a potential new client who is looking for a redesign of their current website, and work alongside us as we move through the 30-step rollout process—all the way through to launching the site. Learn how the latest website development techniques—*responsive design* and *mobile first*—can be woven into traditional project management practices, allowing us to stay current and flexible, while still keeping a rigid focus on planning and communication.

Chapter 1 starts with a review of the resource titles and agency types that we'll come in contact with. It sets up our project with an incoming phone call from Jetzen Spaceships, then plots our course with an overview of the 30 steps. After the first chapter, we'll go through the steps just as we would for a real-life project, with all the required documentation having been created for and applied to Jetzen's responsive website.

Feel like building a website? Come along! I look forward to working with you!

This book has free material available for download from the
Web Added Value™ resource center at *www.jrosspub.com*

1

BASIC TRAINING

If you have a job in the advertising or marketing field you probably work at an agency that produces deliverables that fall into one of these four categories:

1. **Printed material** (catalogs, brochures, posters, manuals, magazine ads, newspaper ads, billboards, etc.)
2. **Digital** (websites, apps, etc.)
3. **Broadcast** (videos, radio spots, television ads, etc.)
4. **Events** (workshops, business shows or meetings, press events, parties, festivals, etc.)

Of course, the list of project types could go on and on, but anything you're working on probably falls into one of those categories. And, the categories often overlap each other. For instance, the event you are managing may require a video or some printed support materials. There may also be a website for registration and event information. This book will focus on managing digital projects. We all know that the technical world moves quite rapidly. Just as soon as we figure out how to manage a certain type of digital project, the technology changes and we're left figuring out an entirely new project. But having worked in this field for several years, I can tell you with certainty that the process doesn't really change—just the projects. The stages of development are always the same, although you might apply them to different technologies and outcomes. For instance, the process I'm going to describe was established before we had to worry about tablet or mobile devices, but I've been able to easily adapt the process to include them.

TYPES OF RESOURCES

If you're just starting out in your career, one of the first things you've probably wondered while checking out the office may be, "Who are all these people, and what do they do?" Believe it or not, most of the project categories listed above use several of the same types of resources. For example, most of them are going to need a writer, an art director, and a project manager. So, before we get into the details of how to manage a digital project, let's get our bearings first. Here's a list of popular resource titles found at a digital agency, and a very brief explanation of what they do.

Brand project manager or account lead: The brand project manager (PM) is usually assigned to a certain brand full time and knows the clients and their business inside and out. When building the business requirements for a project, the brand PM is a good resource for determining what the client's major concerns are, how they'll respond to certain situations or obstacles, and how to best work with that client and their team.

Technical project manager: Tech PMs often get shifted from project to project and are not usually responsible for sustaining or maintaining any one particular brand. During the development process the tech PM is the lead and is often seen as the bridge between the brand team and the tech team (developers, infrastructure, etc.). On my team we call this person the *rollout manager*. They need to understand every facet of the project from both sides of the fence—business and technical.

Business analyst: This role is becoming more and more popular and defined. Not all companies have business analysts on staff because they look to the technical project managers to handle the role of assessing the situation and documenting the requirements. But a more advanced company will invest in a qualified business analyst (BA) to focus on requirements while leaving the solutions and implementation up to the tech or rollout PM. BAs are good at finding a way to explain technical concepts to the clients and brand team in a way they can easily understand.

Asset manager: This is a key role on the production team. The asset manager receives delivery of assets (images, videos, copy, etc.) from the creative team and makes sure they match the specifications of what was expected. This job may also involve searching for available images that fit criteria provided by the art director. In addition, it usually involves keeping track of the assets—which assets are secured, which ones we're still looking for, what was received but did not pass quality review, what retouching is needed, and so forth.

Content authors: Some projects use a content management system (CMS) in place of building html sites. Content authors are experts in how to use the CMS to build websites. Content management systems are most often used for large corporations that want to build one main brand website and then duplicate and localize that site in multiple countries and languages. This is where the term *rollout manager* first evolved, by the way—we build it once and then roll it out around the world.

Art director: These creative individuals determine what the screens will look like in terms of colors, fonts, and layout. They keep busy designing the look and feel for anything that comes out of an agency, such as websites, event promotions, email campaigns, or online banner ads. Or, they may be lending their creative minds to research or brainstorming activities.

Copy writer: Not only do they come up with the headlines and copy, they usually come up with the communication strategy as well.

Information architect: IAs are part designer and part strategist. Through research and focus groups they become experts in consumer online habits and patterns. They weigh the client's business objectives and the creative team's designs along with their own recommendations for a unique and engaging customer experience. This is where science and art come together!

Web developer: Although there's a large variety of programming languages, it's safe to divide developers into two main categories; front-end or web programming. Front-end developers focus on the code that brings the art director's vision to life on screen, while web programmers focus on framework and functionality.

Data architect: This is a type of developer who designs and manages the flow of data from end-to-end. Simply put, if the project requires either a data source or data output, the architect will figure out how to pull the data in, filter it, merge it, present it, and/or push it out to a third party.

Search and analytics: Two groups of resources are included in this category—those who specialize in *search engine optimization* (getting people to visit your website) and *analytics* (professionals who track where visitors came from and what they did while they were there—which buttons did they click on, how long did they spend on a page, and much more).

Infrastructure manager: This important resource manages the hardware and operating systems behind the web services. Servers, disc space, security, and other back-end services are all part of their *domain*. They also administer the different environments available to the developers and user groups. For instance, initial development will normally be done in a *lower* environment, while the testing is done in a *preproduction* environment, and finally released into the *production* environment.

Quality assurance: The quality team is also called the testing team. Once a project is deemed ready, this team takes it and tries to break it. Not *really*, but they do test it against the original business requirements and ensure that the team produced what the client is expecting. They need to test the same experience in all sorts of different browsers and devices.

This is a pretty short list of resource titles found around the typical agency—there are many more! This should, however, provide a fairly accurate list of who's in the room for a digital project.

TYPES OF AGENCIES

There are also many different types of digital agencies:

Digital production house: This is an agency that does nothing but production work, which is rare. The people at this agency do not create assets, and normally have no art directors or copy writers on staff. A production house (or team) takes delivery of the assets from the creative agency and employs a team of developers and content authors who code and produce the end deliverable. Many times the production agency has many other capabilities, it's just that the clients only want to pay for and use the production resources.

Digital creative agency: This is the team of strategists, art directors, and copy writers who decide what the campaign will be, what the desired outcome is, and what it looks like. They employ art directors and copy writers to hand over design files to the production team for execution and delivery.

Social media: Facebook, Twitter, Snapchat, Instagram, Vine, etc.—there are so many different options for social media, and this team recommends the best platform on which to deliver a client's campaign.

They strategize on how to best represent the brand on various media outlets in order to achieve the marketing and advertising goals.

Search engine optimization (SEO) agencies: SEO professionals know how to direct Google and Yahoo users to your website or application. Even though PMs aren't personally responsible for SEO, it's one of the most important things we can do to ensure a project's success, thus, we'll be talking a lot about this category. Why create a website if nobody can find it?

Analytics: A wise man once said to me, "If you can't measure it, you can't manage it, and you shouldn't be doing it." This applies to many aspects of advertising and marketing, and exceedingly so for websites. If we're being paid to produce a website or application, it's probably business-related and the clients have bottom-line results in mind—retail results.

Full-service digital agency: A full-service agency can do everything from television advertising to events and printed materials, and for digital this means they have every category previously listed in one place.

A client could hire each of these agencies individually or they could hire a full-service agency that does it all. But for the purposes of this book, let's say we are PMs working at a digital production house, and the project is to build a responsive website using a mobile first methodology.

RESPONSIVE DESIGN

Not too long ago there were plenty of agencies who would brand and market themselves as mobile application specialists, and that was fine because it was a new platform, a new technology, and we were all still learning how to adapt to it. But these days we have to do it all—desktop, tablet, mobile, and whatever comes next. It's just not acceptable to build a digital project that isn't responsive.

A responsive website is designed to provide an optimal viewing experience no matter what the screen size of the end user. Not only does it need to have a fluid layout and navigation which resizes and repositions itself depending on screen size, it has to also provide a user experience that's both unique and consistent at the same time. Unique in that with a mobile-sized screen we just can't provide the exact same view that one would see on a larger desktop screen. That would be senseless, the screen is too small for that to work. But, it also has to be consistent because customers shop between devices and we don't want to confuse or agitate them with a mobile phone version of the website that's completely different than what they saw using their desktop computer. Test this out for yourself by visiting the same website on your desktop computer, smartphone, and tablet. Unless you have a huge desktop screen the experience will probably match the tablet version, but I hope the mobile version is different (but similar).

MOBILE FIRST

There are a couple of different methods for how to approach responsive design. Historically, because the desktop came first, mobile was sort of an afterthought to the desktop experience. In other words, website designers would create the optimal desktop experience without much consideration for the mobile users, and then afterward try to jimmy the desktop experience down to size. This led to a period of time where it was considered best practice to make two completely different websites for the same client. So, we'd build them one site to cater to desktop users and then a whole new site which would only be shown on mobile devices. Internet magic has something called *mobile detection* which is basically wizards living in the internet who can tell if the viewer is using a small screen and if so, send them to the small screen version of the website. Anyway, as you can imagine, the practice of building two different websites was expensive, time consuming, and irritating because *just* when we thought we were done, we had to start all over again.

Also during this time period, mobile bandwidth globally was very inconsistent. In the United States and other first world nations it was *anything goes*—we could stream videos and load content quite quickly. The content side was actually trying to keep up with everything the technology side was allowing us to do. But in other parts of the world, even as close as South America, bandwidth was just not there yet and they couldn't handle the same experience. Low bandwidth meant slow page-load times and almost completely ruled out videos. Not to mention that smartphones weren't even readily available in those countries, so we actually had to make a third type of experience for any one website. *Three* different sites? Yeah—the desktop version, the high-fidelity U.S. mobile version, and the low- or medium-fidelity mobile version for low-bandwidth countries.

Today, technology has improved. Low-bandwidth countries are catching up with high-bandwidth ones, and clients everywhere are demanding responsive design. And I don't disagree—we should build one site that adapts (or is responsive) to the viewing environment using proportion-based grids, flexible images, and an altered CSS (Cascading Style Sheet). I really can't explain CSS right now but will attempt to later in the book. But here's the question: do we start with the mobile screen in mind and then expand it to fit larger screens (mobile first), or do we create the optimal desktop experience and then size it down? Well, that's debatable, and I'm not even sure there's one correct answer to that question—it depends on the project.

The reason so many people like the concept of *mobile first* is because if we try to design the desktop version first and then scale down, it's quite difficult. It's a lot harder to start with robust material and figure out what to delete than to start with a small screen in mind and then add to it. I like this approach because it forces us to come up with a clean, clear design which is easy to follow and navigate.

Another more obvious way to solve this problem is to consider the original concept of the project. Of course, if we're building an experience that mostly benefits a person who's in the middle of a big city with only a phone in their hand, we go mobile first and then make sure that the desktop site looks good as well. A good example for this might be a mobile app for ordering car service. People also need to order car service from their homes, but it seems like one shouldn't have to turn on a computer for something like that.

On the contrary there are lots of other websites more fitting for a larger screen, such as financial programs or large purchase items, just to name a couple. Experts don't agree, however. Some say that *all* websites should be developed as responsive, and others say that there are some user experiences that will never work well on a small screen. As PMs, however, this is not our battle. We have other experts in the room who make these types of decisions—people whose job it is to make these calls. Our job is to create and manage the process and keep everyone on task. For the purposes of this book, we're going with mobile first. We'll design first for the mobile users and then expand for desktop.

The process I'm about to describe works for any type of digital project, whether it be a tablet application purchased through the iPad store or a global web experience on all types of devices in all types of languages. How could that be? It's because I'm not explaining how to develop the code or write the analytics—I'm just providing a list of line items we need to hit for each project. Every project requires code development, and the code will change based on the project. That's an *NMP* in my world—not my problem. (Don't say that out loud.)

I think the best thing to do at this point is to just get on with it so you can see for yourself. Let's make a website! Ready?

GETTING THE ASSIGNMENT

(Office phone rings)

[Project manager] "Hello?"

[Account manager] "Hey bud, how ya doin'?"

[PM] "Great. It's a little cold in here today, but great otherwise. How about you?"

[Account manager] "I'm doing great, too. Say, we have new a client interested in hiring us to build a website for selling spaceships, and we need to get back to them with an estimate on time and costs. Do you have something *in the can* that you can send me?"

[PM] "Well, I'd have to know a lot more about the project before I can provide any estimates. These types of projects can vary widely, as you can imagine. We actually have a rollout process that we follow that starts with an initiation phase where we nail down the business requirements and then provide the estimate."

[Account manager] "Right, and we definitely want to follow the process and do this thing the right way, but for now we just need ballparks. They just want to know what they're getting into."

[PM] "Hmmm. Well, can you think of a website we've built in the past that this would compare to?"

[Account manager] "Not really, and that's part of the problem. The client doesn't even know what they want. We need to guide them. So, we thought we'd send them a ballpark estimate along with a generic schedule so we could start to educate them on the process and what's all involved with this."

I actually get this call all the time. Do you? If you've been doing this for any length of time, I bet you've fielded a few phone calls like this. The problem here is clear, right? How can we provide a budget for a project we know almost nothing about? And why are we the only ones who understand that this is a problem? Well, the sales team doesn't always seem to get it. But, we probably don't fully understand their worlds either, so let's be nice.

First, remember, you're a professional PM and also a process consultant. When somebody asks you this type of question it's clear that they don't really know what they're asking for any more than their clients do. In my opinion, they may not be verbally asking for my help, but they sure do need it. So my response would be something like this:

[PM] "This project sounds really exciting and interesting—I love spaceships! I think we could send them high-level estimates for building a generic small, medium, or large website, but also explain to them how vastly different the costs can be depending on the business and functional requirements. Make sure they understand that these are just ballpark figures…."

[Account manager] "Oh, of course."

[PM] "… and that we would love to meet with them to review the rollout process we follow. I have a process map I can walk them through, and I can put together some questions that would help me better understand what they're asking for—we should get the IAs on the call as well. Actually, I'd be happy to meet with the internal team first to review the process before showing it to the client so that we're all on the same page. How does that sound?"

With a quick sentence or two I was able to introduce leadership, instill confidence, and initiate a process.

Introduce leadership: Instead of just sending over what the account manager asked for while my brain slowly melted and peeled away from frustration, I decided to change the paradigm. Sending random costs and timelines over would result in our own team selling something we could never deliver on. Once we send over costs and timelines to people, that *is* what they're going to sell and then try to hold us to. Believe me, that will come back to bite us. Instead, I flipped it around and introduced leadership. I met them halfway by sending along some size-based budgets which could cover any range of project scopes, and offered to lead the team by teaching them the process—my process. You can cover the budget estimates by sending over three somewhat random, but also logical, cost breakdowns—something like this:

Small = $50k–$125K
Medium = $125K–$500K
Large = $500K+

It could be anything—work with the sales people on this—it's really more their job than a PM's at this point.

Instill confidence: Through offering to lead the team and teach the process, I instilled confidence by letting the account manager know that they've got somebody on the team (me) who can speak intelligently to the client. Why wouldn't they want to show that they have a defined digital process and a master curator of such? To be truthful, if the clients wanted to figure this stuff out for themselves, they would have—they want to pay us to know how to do it.

Initiate a process: Inserting the process at the very beginning gives us control before things turn to disaster. Here's a circumstance I've come across time and time again: the sales team rushes out on their own without subject matter experts to help consult, and promise everything under the moon to the clients. Then they hand that huge ball of mess over to the PM to deliver. "Wait…we can't do this…especially not for that price.…" Ever been there?

In another scenario, I've seen sales teams sell multimillion-dollar programs without any detail whatsoever about what the deliverables will be. This situation isn't as bad, because you still have half a chance of putting together a reasonable scope. But I'm always amazed when clients sign a proposal for unidentified deliverables. I'm both amazed at the sales team for being able to pull that off, and also amazed at the purchasing departments who allow it. Last time this happened to me, the CEO of a company I was working for gave me this huge book of a proposal that was sold to a particular client and told me to "git r done." After reading through the proposal I was really confused and told the CEO that it didn't seem to contain any actual deliverables. He just smiled, gave me a wink, and walked away.

THE ROLLOUT PROCESS

Now that we have our project identified (making a website of some sort), let's call the make-believe company *Jetzen Spaceships*. We really don't have any information at this point, so what should we do first? The process I'm about to explain has been tested, refined, and tested again. I've been developing this process for years and can confidently proclaim that as a practitioner I refer to it for every single digital project to which I am assigned as PM or consultant. It won't let you down.

Although the process aligns with the process groups found in the PMI® *Project Management Body of Knowledge* (*PMBOK® Guide*), all of these steps are grouped into two main categories—*Plan and Define* and *Construct to Close*. I felt it was important to break up Plan and Define into two areas (Initiating and Planning) because there is a clear tollgate after initiation where the client reviews and approves the project scope and budget. Other than that, the appropriate digital development steps are listed either under Plan and Define or Construct to Close. Table 1.1 shows the 30 rollout steps.

I'm a big believer in planning and have witnessed a huge difference in the success of projects where there was a lot of time spent up-front planning things out rather than just going with the flow. The bulk of the work is always in the planning column.

Although the steps are numbered, they're not always done in succession, and that's important to note because when showing all 30 steps to the team, it's a good idea to show the graphic as well (Table 1.2) or risk having the account team and clients freak out at the sheer number of steps. It just *looks* like a long process—but, as you can see from the Gantt chart, these steps often overlap, and when clients see this graphic they're not only assured that the team knows what they're doing, but also that they're doing everything possible to move quickly.

Notice that not all of the 30 steps are included in the Gantt chart—I don't call out things like communication planning or analytics and SEO analysis. These things are not part of the critical path and just need to get done in tandem with the rest of the steps. There are a lot of steps because PMs need to be detailed so we don't miss anything—but really, it boils down to about 11 key milestones. Once we show a chart like this to the team, they tend to calm down and get onboard. You might not even want to show them the entire list of 30 steps—that's up to you.

In addition, development can be in either Plan and Define or Construct to Close. In Plan and Define, we definitely figure out what each of our development tasks are and get them priced out and scheduled, but we don't necessarily start the actual programming unless the clients have already agreed to pay for it. We'll go into more detail about this

Table 1.1   Rollout process

PLAN AND DEFINE	
INITIATING	**PLANNING**
1. Gap Analysis	7. Plan and Define Schedule
2. Workshop	8. Communication Planning
3. Stakeholder List	9. Sitemap
4. Business Requirements Document	10. Technical Solution Strategy
5. Preliminary Budget Estimate	11. Wireframes and Functional Specifications
6. Statement of Work **(Tollgate)**	12. Styleguide
	13. Analytics Analysis
	14. SEO Analysis
	15. Infrastructure Assessment
	16. IT BOM
	17. Development and Change Management
	18. Test Strategy and Cases
	19. Package Identification
	20. Construct to Close Schedule
CONSTRUCT TO CLOSE	
21. Create Content Tracker	
22. Asset Quality Review	
23. Content Entry	
24. Quality Assurance	
25. System Integration Testing (SIT)	
26. User Acceptance Testing (UAT)	
27. Nonfunctional Testing (Performance, Security, Disaster Recovery, Failover)	
28. 301 Redirects	
29. Cutover Management	
30. Transition to Operations	

later, but for now, take a look at the items that are highlighted in bold in Table 1.2.

Budget approval—This is important for two reasons: first, the clients need to know when they can expect to receive a budget, and second, we need to point out that work will not proceed past this point until the budget is approved.

Table 1.2 Use this Gantt chart to explain how the tasks often overlap each other.

		Month 1	Month 2	Month 3	Month 4	Month 5
Plan and Define	Gap Analysis	▬				
	Needs Asmt.	▬				
	Budget Approval		▪			
	Sitemap		▪			
	Technical Solution		▬			
	Wireframes			▬		
	Styling			▬		
Construct-to-Close	Development				▬	
	Content Entry				▬	
	Testing				▬	
	Cutover					▪

Content entry—This is important because it also affects the creative team. They need to know at what point assets are expected to be delivered so they can allocate resources and get their own budgets approved.

Another thing the clients like to hear at this point is that the team will have a *transition-to-operations* period immediately following cutover. This is generally a two-week period where the build team stays on the project and handles any lingering issues or bugs, and are also there for support, just in case something unexpected happens. Once the site is considered stable, I always prepare some training for the sustain team. (More on this near the end of the book.)

Of course, these steps may change based on the nature of the project, but they're an excellent baseline from which to begin the assessment. Remember also that these steps can be applied to any digital project, whether it's a website, application, or mobile tool. As long as it's a consumer-facing project that requires some sort of cutover activity.

Starting with the Initiation Phase we're going to learn step-by-step how to approach a digital project using Jetzen Spaceships as our example project. It would probably help us to understand a little more about Jetzen before we dive in. Since they're make-believe, it may be difficult to conceptualize what we're doing and why, so here's a make-believe profile on them:

Jetzen Spaceships is a global manufacturer and seller of several different models of spaceships for both public and corporate use.

Headquarters:	Marana, Arizona
U.S. Zone Offices:	Richmond, VA
	Newton, KS
	Novato, CA
	Syracuse, NY
Global Offices:	Mannheim, Germany
	Shanghai, China
No. of Employees:	19,273
Sector:	Manufacturing
Sales:	$56.4 Billion 2014

Finally, I'd like to show you how I've organized this book in terms of chapters compared to the process (see Table 1.3). Take a look before we jump right into the initiation phase for our new client, Jetzen.

Let's make a website!

Table 1.3 Contents of book

Chapter	Process Steps Covered
Introduction	
Basic Training	
Needs Assessment	1. Gap Analysis 2. Workshop 3. Stakeholder List
Documentation	4. Business Requirements Document 5. Preliminary Budget Estimate 6. Statement of Work
Communication	7. Plan and Define Schedule 8. Communication Planning
Information Architecture	9. Sitemap 10. Technical Solution Strategy 11. Wireframes and Functional Specifications 12. Styleguide
Analysis	13. Analytics Analysis 14. SEO Analysis 15. Infrastructure Assessment 16. IT BOM
Development	17. Development and Change Management
Preparation	18. Test Strategy and Cases 19. Package Identification 20. Construct to Close Schedule
Content Input	21. Create Content Tracker 22. Asset Quality Review 23. Content Entry 24. Quality Analysis
Testing	25. System Integration Testing (SIT) 26. User Acceptance Testing (UAT) 27. Nonfunctional Testing (Performance, Security, Disaster Recovery, Failover)
Cutover	28. 301 Redirects 29. Cutover Management 30. Transition to Operations

SECTION I: INITIATING

2

NEEDS ASSESSMENT

Chapter one was just a set-up for our project. Now we're going to start learning the rollout process and following the steps. Assignments come in to an agency through several different channels. One example is a request for proposal (RFP), also referred to as a request for quotation (RFQ). They're usually sent out by the purchasing departments of the requesting company and contain detailed information about the project requirements and deliverables. Normally RFPs or RFQs are sent out to a select list of potential suppliers who are given time for questions and answers once the agency has had time to read through and digest the written material. This is an excellent way of doing things because it provides quality communications and information to all parties right up front.

Other times, work comes in through a phone call or hallway conversation—like the one described in the previous chapter. If that happens, we have to figure out for ourselves what the clients are looking for because they may not fully understand the assignment themselves. The call probably came in from the marketing or brand client who received the assignment from their boss, and now they're flipping it over to an agency to come up with a response and plan of action.

Most companies today have some sort of web presence—especially large companies. I would think a spaceship company would have something online, but you never know, right? At the beginning of the project we always have so many questions:

1. How big will it be?
2. What type of applications or tools do they need?

3. Who will be providing the assets?
4. What is the URL for their current site?
5. When is their target launch date?
6. Will they let us test drive a spaceship???

This list goes on and on and I always refer to this as *initial research*—not too complicated. Our first responsibility is to develop a project charter to which the clients can react. Sometimes called a *statement of work*, this document details exactly what the request is so that there's no confusion down the road if somebody tries to increase scope or challenge costs. Three main documents feed into the statement of work:

1. Gap analysis
2. Business requirements document
3. Budget estimate

But before we go any further, we need to agree on some basic project assumptions since this is a make-believe situation. Let's say that after those basic budget parameters were sent over to the new client (based on that crazy phone call that came in from the account manager), the client replied by sending back some comps that their creative agency produced for a website redesign. Now, as the saying goes, we're cooking with gas.

GAP ANALYSIS

Let's start by reviewing their current website versus the comps and start putting together a list of differences, or *gaps*. Creating this list is important for several reasons:

1. Jetzen Spaceships already has a database solution for their existing website functionality—they'd have to in order to support the current website's functionality. But they might not realize that the newly proposed design from the creative team requires data that isn't currently supported or even available. Later on in the process we'll create a technical solution strategy which will specify if and how the existing database needs to be adjusted. This may create costs for Jetzen that they hadn't planned for.
2. The new design may not include all of the functionality of the current website, and this may very well have been an oversight.

It's possible that the creative team missed some pages during their review, so we should do a meticulous page-by-page audit just to make sure everything has been accounted for.

3. Whatever tools currently exist on the site will have to be rebuilt by our team because the current design isn't responsive and looks totally different than what we have planned. This list of *requirements* will feed into the business requirements document that we have to complete next, which feeds into the statement of work, so it has to be accurate.

What do you think of when you hear the word *gap*? When I think of this word, I picture a big hole in the ground with the client's requests on one side, what the agency wants to deliver on the other side, and all the loose parts falling into the gap in the middle. Well, let's get this negative word association out of our minds right now. Focus more on the word *analysis* because, in fact, we are studying the project at this point. If there's a business analyst on your team they might call this the *needs assessment*, and we would be developing the *situation statement*.

There are a couple of different ways to look at the gap analysis. For the *Jetzen* project we're going to compare the current site to the new design and see if we find any gaps, as mentioned before. Any discrepancies will be highlighted and shared during a workshop we're soon going to host so we can get answers to anything not completely understood. Already we know that the original site was not designed as a responsive solution, and that particular part of our recommended solution is going to have a huge impact on their business.

But what if there isn't a current site? Can we still call it a gap analysis if the analysis is based on a brand new design or idea? Sure we can. In this case we read through all of the available material (design comps, RFP, meeting notes, sketch, etc.) and anything we don't understand or feel may be challenging is a gap. On the next page is a good graphic (Figure 2.1) that someone drew for me at one point in my career, and I think it applies to most project situations and certainly to a gap analysis.

The gap analysis should be performed by someone who is able to ascertain how a site is currently structured (sitemap), what the current customer experience is, and which, if any, designs require new technical solutions to be built in order to achieve the desired outcomes. The appropriate resource for this analysis is an information architect (IA)—and I usually give the IA at least a week or two to complete this task, depending on their workload.

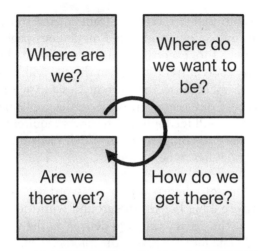

Figure 2.1   The gap analysis falls between the upper two boxes: Where are we? and Where do we want to be?

Again, the first step is for the IA to take a good look at the current website and take notes on how it currently operates and functions. From this he or she will create an *as-is* version of the sitemap so we can compare it to the new one we're recommending. Looking at the current website (Figure 2.2), it doesn't seem very well organized and has little to offer the customer. It's more like an online brochure. But it does have photos, news articles, links to social media, and video in addition to some basic product information. During the analysis it's prudent to jot down a list of questions to ask the client. Here are just a few examples:

1. How are the news stories fed? Are they automatically updated through a feed or do you update the content manually?
2. How often do the news stories update, or how often do you have new stories to add?
3. Same questions for the promotional offers and events.
4. What is your social media strategy?
5. Where are your videos hosted?
6. Careers seems to link to an external site—do you want to keep that link-out or would you like careers built into the new website?

Once we have a good idea of how the current site operates we can review the redesign comps and start to list out the gaps. During our

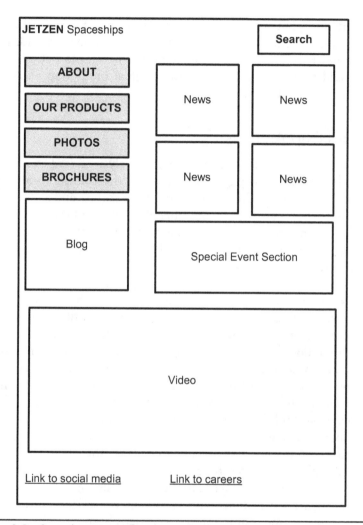

Figure 2.2 Start the gap analysis by analyzing the current site

review of the comps we see that there's now a store locator tool, and a new *contact us* form. The clients have also mentioned that the spaceship buying process is very complicated and customers often feel overwhelmed by how many choices they face. They asked for ideas on how the process could be simplified or if there is a better way to present the material to make it easier to understand. Above all else, the client wants to increase sales. All of this information goes into the gap analysis presentation, slide-by-slide.

No matter what we're starting with—whether it's an existing website, comps, sketches, or even written documentation—the idea came from somewhere. If we were trying to build a better mousetrap, the first step would be to figure out how the original mousetrap worked and strategize on how to make it better.

Jetzen's Gap Analysis

When the IA is ready to present their gap analysis slides I expect to see screen shots from most pages or sections of the current website along with any new functionality coming out of the comps, and also any other ideas the IA has and thinks would benefit this client. Sometimes they pull ideas from other *benchmark* projects. Based on the information the clients gave us up front along with the IA's research, we might get something like this:

- **News stories** are currently updated manually, but the clients have a media RSS feed that they would like to utilize to provide a timely refresh of new stories throughout the week.
- **Promotions** are only offered nationally, are not based on customer location, and are updated about every other month. A manual process is fine for these updates since the workload is minimal.
- **Events** information is often updated and would benefit from a calendar tool that can be sorted by location.
- **Social engagements** are a key part of Jetzen's online communication strategy and they'd like to increase their brand's awareness through visibility on these sites.
- **Videos** are not considered in the current site or on the comps, but through conversations with the client we've learned that Jetzen has an enormous amount of cool videos hosted on their YouTube channel. This would be a great addition to the site.
- **Careers** will have to continue to be hosted outside of our website due to Jetzen company firewall restrictions.
- A **store locator tool** should be added to lead customers into their local stores.
- **An interactive spaceship configurator** would help the client provide a better online shopping experience for the customers. The current site doesn't have a database to support a configurator—we'd need to either figure out if this tool can be supported by an xml file or if we can tap into the company's parts and packaging ordering systems for support.

By reviewing these examples, hopefully it's clear that at this point it's not necessary to explain how to achieve these goals, just that they need to be accounted for. Now that we have the gap analysis ready, it's time to present it to everybody at an all-team workshop. Throughout the presentation, open dialogue will be encouraged so we can capture not only all of the requirements for this project, but also any other potential gaps that we haven't yet accounted for. The purpose of the presentation is to make sure all parties (client, creative, production, etc.) agree on what is being built. This will lead us to business requirements, budget, and schedule.

CONDUCT THE WORKSHOP

Now let's start planning for our first all-team workshop. Here are the major points I normally cover during the initial workshop:

- Business lead (client) name and introduction
- Agency partner names, roles, and responsibilities
- Project name and objectives
- Review of comps or initial requirements
- Gap analysis presentation
- Confirmation of next steps

The purpose of the workshop is to indoctrinate all parties into the project, communicate the client's expectations to the team, and allow them to provide input to the gap analysis. By the end of this workshop we should be able to move forward with gusto. It's also an opportunity to establish ourselves as the project lead—the rollout manager. Work to completely understand the project without having to rely on the IA to answer all the questions. This may be difficult at first, but once you've been through a few of these you'll soon be an expert as long as you take the time to question and understand everything that's going on. If we can't provide any answers, the team won't ever look to us as their leader.

It's important to schedule the workshop during a time when most participants can attend since this is really the kickoff for the project. But, who should be invited? A big component of project management is communication and we don't want anybody to feel like they weren't given a heads-up or opportunity to provide input on the project, so invite everybody and let them each decide whether or not they'd like to attend. To start an invitation list simply refer back to the resource list:

- Brand project manager or account lead
- Technical or rollout project manager (this might be you)
- Asset manager or content lead
- Art director
- Copy writer
- Information architect
- Data architect
- Developer(s)
- Search and analytics
- Infrastructure manager
- Quality assurance

Don't forget the clients and any other agencies involved! Since this is the first time many of these folks will have heard of the project, we may need to contact their functional managers for resource assignments. Let's go through the agenda items and discuss how we'll review each of these areas:

- **Business lead (client) name and introduction:** Always make sure to introduce the client first. Most people *should* already know who the client is, but it's important to make the client feel like they're the most important person in the room—because they are.
- **Agency partner names, roles, and responsibilities:** Go around the room and ask everybody to introduce themselves—stating their name, which company they work for, and what they do. Make sure to take notes during this because we'll need to start a *stakeholder list* right after the meeting.
- **Project name and objectives:** Start with a background of the project—perhaps the client can do this part. Remember that some people in the room know almost nothing about the project, so it's a good idea to review how the project was initiated and what the objectives are. For instance, for the Jetzen project, the client might say something like this: "We've had our current website up for ten years now and we're ready for an updated customer experience. The goal is to create a world-class, responsive website that houses all of our product information, specifications, corporate

information, and purchasing tools. We've shared some new design concepts with the production agency and are ready to move forward." The clients may or may not mention *mobile first*. This is a rather new concept that the agency may end up bringing to their attention.

- **Review of comps or initial requirements:** Inputs for this section can be the RFP, comps, meeting notes, or whatever else might be needed to explain the project. Workshops are meant to be working meetings, so it's not as important to have a polished presentation as it is to make sure that every detail is covered. Open up the floor to questions throughout the process—all of the experts are now in the room and this is the best time to get their input.

- **Gap analysis presentation:** The gap analysis is going to be an excellent start to the project, but remember that it was put together in a bit of a vacuum, so be prepared for input. For instance, the search engine optimization (SEO) team may mention an up-coming URL strategy that the team needs to incorporate into the requirements. Or, the analytics team may need certain types of tracking on the new site that's considered up-and-above traditional web analytics. Keep track of all of the functionality requirements throughout the discussion and start marking them down. At some point we'll need to deliver a business requirements document (BRD) that contains complete details on the functionality of each requirement. The BRD needs to be detailed enough for the developers and IA to work from, in order to produce an acceptable final product—and the quality team will also use this document to produce their test cases and scripts.

- **Confirm next steps:** The team will want to know what the next steps are, now that the project has been kicked off. The next step might be that the team needs to attend a second or even third workshop in order to cover all open items. If we already have all of the requirements defined, tell the team that meeting notes will be distributed and the next step will be to review and discuss the BRD.

That wasn't so hard, was it? Outputs from the workshop are meeting notes, a stakeholder list, and at least the start of a BRD. I go into more detail about the stakeholder list and BRD shortly.

THE STAKEHOLDER LIST

One of the first things to tackle after the workshop is to make sure we have a list of everybody involved in the project, and what they do. This stakeholder list is a very important part of the process and I find myself referring to it often. It should list every client, art director, developer, and anybody else who will be touching the project. Table 2.1 shows an example of a stakeholder list. I don't normally put a ton of information in mine. For instance, I don't usually list phone numbers because I rarely call people. I e-mail them. But, like all example documents in this book, you'll be the owner and can adjust them as needed or desired.

People may question why I even include something as simple as a *contact list* like this as part of my key process steps. I find it to be incredibly important, and something that many people skip. Following are a few reasons I believe the stakeholder list is so valuable:

My poor memory: In addition to the resources already named in this book, there'll be many more added during the life cycle of the project—several of whom will be random people associated with the client

Table 2.1   Stakeholder list

STAKEHOLDER LIST			
Name	Company	Role	E-Mail
		Content Delivery Network	
		Brand Project Manager	
		Tech Project Manager	
		QA Lead	
		Client Brand Lead	
		Client Tech Lead	
		Analytics	
		SEO	
		Creative Director	
		Senior Designer	
		Copy Writer	
		Information Architect	
		Infrastructure	
		Host Provider	
		Application Developer	
		Java Developer	
		CSS Developer	
		Data Achitect	

and who *own* different sections of content or data. Add them to the list as we go so we can refer back and don't have to ask twice whom the correct parties are.

Cutover activities: Near the end of the project everything is tested thoroughly before and after it launches. The last thing we need is somebody coming forward after launch saying that they were never contacted to approve their section. I like to keep everybody on the list updated regarding project status, development, and launch activities—just in case they have something to add. That way, I'm covered if anybody isn't happy with something that goes live.

Client service: Stakeholder lists are also very helpful to the client. They might not be comfortable broadcasting that they don't know who to contact for certain things. This way they can be discreet and look at the list. Stakeholder lists also make us look super buttoned-up as project managers.

Sustain: After cutover, the build team hands off operations to the *sustain* or *maintenance* team. The stakeholder list will be helpful to them because they weren't involved during the build and may have no idea who to contact if something goes wrong.

Once we have the stakeholder list complete, or at least populated with the key people from the workshop, be sure to distribute it to everybody on the list. Once I was training a new project manager (PM) and teaching her how to make project plans and contact lists and so forth, and she was doing a great job at updating them throughout the project. I was actually new at training PMs and was thinking to myself that I was doing a pretty good job. Then, at some point she didn't get a deliverable on time from one of the functional groups and was wondering out loud why they didn't send their materials. I asked if she thought they were using her project plan and she said, "Oh, I was supposed to send that out?" Well, yeah. I didn't think I had to specify that—but I guess I did! So much for me doing such a bang-up job.

At this point, we should have all of the information we need to start building our documentation, which I'll cover in the next chapter.

In this chapter we:

- ✔ Did some initial research
- ✔ Initiated the gap analysis
- ✔ Held a workshop
- ✔ Created the stakeholder list

3

DOCUMENTATION

This section takes us through all of the documentation that's not necessarily required, but good practice for any project. The purpose of creating solid documentation around deliverables and project activities is to establish clear objectives, goals, and acceptance criteria for completion. Also, in the digital world it helps the production team get the details they need surrounding functionality. Business analysts (BAs) will sometimes fulfill the role of documenter, but it can also be done by the tech project manager (PM) or rollout PM. We begin with the business requirements document (BRD).

BUSINESS REQUIREMENTS DOCUMENT

The BRD provides detailed definitions about the requirements so that the team has a complete understanding of what the final product(s) should be. There are three types of requirements—*business, functional,* and *technical.*

Business requirements: The business requirements usually come from either the client or the creative agency. Think of it as a general description of *what* needs to be achieved, not *how* to achieve it. Let's take the Jetzen news stories requirement as an example. The business requirement might be something like this:

- News stories should be fed from the Jetzen Media Department's RSS feed and updated every two hours as stories are received

Notice, not a lot of detail is needed for the business requirement.

Functional requirements: Functional requirements give the detail behind the requirements. What happens when you hit the submit button? Where will the form be sent once the customer fills it out? How should the news stories be filtered and presented on the page?

For example:

- News stories should be sorted into at least one of these five categories:
  - ▼ Awards
  - ▼ Community
  - ▼ Mission blogs
  - ▼ Technology
  - ▼ Events
- One news story may appear in up to two categories
- The News Stories page must be refreshed and cached every two hours
- News stories will be automatically pulled in from the Jetzen Media Department RSS feed, and displayed on the web without edit
- The News Stories landing page will list the five main categories, with teasers for each story under the appropriate category
- When users click on a teaser they'll be sent to the *Details* page for the story
- And so on…

If possible show any comps or sketches available to help describe what should be happening on the screen. Be as detailed as possible and tenacious with the descriptions to capture every possible scenario. Some functional requirements describe what happens as a result of user actions, but other requirements describe what that business (client) needs to happen. For instance, describing what happens when the user clicks on a teaser is the result of a user action, but describing what the five main categories need to be is a business objective.

Some business analysts like to write *use cases* in place of functional requirements. And I'm fine with that, although I don't use that technique myself. I'm not going to go into a lot of detail here because if we were using the use case approach then the entire business requirements document would be very different. But just to give an example, a use case for news stories would read something like, "I need to find the latest news stories on Jetzen Spaceships—all in one place." It's an interesting approach, but not one I'm taking in this book.

Technical requirements: Although I'll describe what technical requirements are, I never put them into the BRD because defining them is the job of the data architects and developers. This is the *how* part and I'm not going to try to make something up here. The tech team will have to figure out how to grab that RSS feed, import it, filter it, display it, and cache it. They don't make schedules, and I don't try to figure that stuff out.

How Do You Eat an Elephant?

One bite at a time. Try not to get overwhelmed at the task of creating the BRD. It's a huge document, and yes, it will probably take a long time to create, but just do it one step at a time. I always start by creating the section headers first—one for each piece of development work. For the Jetzen project I created one page for each deliverable, like the example shown in Figure 3.1.

JETZEN BUSINESS REQUIREMENTS DOCUMENT

Jetzen
Business Requirements Document
Table of Contents

Figure 3.1 Start with a list of section headers and begin to jot down requirements throughout the process

The internet has lots of examples of BRD templates from which to choose, if your company doesn't already have one in place. Any format is acceptable, as long as it provides the team with everything they need to understand about what the clients are asking for. It could be done in a spreadsheet, text document, or slides. As I stated before, left-brained people are pretty averse to reading through documentation, so I sometimes have to take the BRD and convert it into slides to make it easier to digest.

The goal is to list every single business requirement for every piece of development work. This document will be signed off on once complete, and the information architect (IA) team will move forward with their job, using it as one of their inputs. With this in mind, please note that no detail is too trivial. My rule of thumb is (and I tell this to my clients), "If it isn't listed in the BRD, don't expect it to be delivered." Please stress this to the team, because, for sure, someone will be testing the final product and come up with something that *they thought was going to be included, but didn't think they had to mention.* Yes, please mention it.

The BRD is also the road map which guides scope management throughout the life cycle of the project. Once we lock down the business requirements, we can then refer back to them if somebody tries to add requirements after scope has been finalized (they will, for sure). It's never a problem if the client wants to add requirements, but another rule of mine is that new requirements need to be officially added to scope, budget, and schedule. What this means is that new requirements may very well cost more money and may not get done at the same time as the original requirements. We may have to add a Phase two. Be diligent about this, or the project will spiral out of control.

I like to start a first draft of the BRD, prior to going into the first workshop—that way I can take notes right inside of it. I begin by making a section header for every known business requirement that was gathered from reading through the gap analysis in advance, then as the gap analysis is being presented and different team members start mentioning specific business or functional requirements, I can easily add them to the document. In addition, we can confirm with the team any requirements that we've been able to discern thus far.

Here's an example of why it's important to talk through every detail. The contact form looks pretty simple, but maybe behind the scenes, the customer information is being sent to different zone offices based on the customer's geographical location or internet protocol (IP) address. This is something that must be covered in the BRD so that the developers can provide the correct solution; we have to find a way of getting these details out in the open. Don't be afraid to ask plenty of questions about functionality and data flow—it won't take long before little unexpected details jump out and surprise you. When this happens I find myself thinking something like, "You didn't think I needed to know that the contact form needs to be sent to the moon for authorization and processing?" That may very well happen (not the moon part, but something like that). Figure 3.2 shows examples of how one might put together a BRD for a particular tool like News Stories.

Each business requirement has definitions for each of these five categories:

1. **Source of requirement:** It's helpful to understand where the requirement is coming from so that the designers can better understand the objectives of the requirement, as well as the level of importance. Remember to collect contact information for the person responsible for each requirement and add them to the stakeholder list.
2. **Business requirement:** This should be a general overview and concise description.
3. **Functional requirement:** The business rules and functional behavior, either on screen or behind the scenes.
4. **Dependencies:** Many times the requirements rely on a third party for data, information, content, or additional development work done outside of the core team. Make sure the client understands that the requirement can't be completed unless the dependencies are achieved. This will also alert the client that they need to make the dependencies a priority for the third party, who may not be as engaged as the core team.
5. **Notes:** This can be anything—just an extra spot to put pertinent information.

The BRD doesn't include every single page or content element that was discussed during the workshop. For instance, I wouldn't mention how the client's promotions are national and not regional, so there's no reason for a special tool to organize and publish them because that wouldn't require special functionality or development. It will simply end up as content that will be provided and uploaded to the site. Now, if we needed to provide a tool whereby the consumer could sort for local promotions, then we would need to add it to the BRD.

Title	**NEWS STORIES**
Source of Requirement	Jetzen Client – Media Relations Team

Business Requirement
News stories should be fed from the Jetzen Media Department's RSS feed and updated every two hours as new stories are received

Functional Requirements
News stories should be sorted into at least one of these five categories 　o　Awards 　o　Community 　o　Mission Blogs 　o　Technology 　o　Events One news story may appear in up to two categories News stories web page must be refreshed and cached every two hours News stories will be automatically pulled in from the Jetzen Media Department RSS feed and displayed on the web without edit The News Stories landing page will contain teasers for each story which will be separated by the five categories When users click on a teaser they will be sent to the detail page for the story upon which they clicked News story copy will be accompanied by one image per story Images will be delivered at 560x250px News section will contain one landing page listing a blurb on all categories and associated stories. Upon clicking on a story title or image, the customer will be directed to a page dedicated to that story The first sentence from each story feed will be used, along with the image, as the blurb for the landing page Most recent stories should be sorted on top

Dependencies
Jetzen media must update RSS feed with stories as they become available

Notes

Figure 3.2　Every topic should have its own page or section, depending on length

Since this is our first job with Jetzen, I also included some documentation (see Figure 3.3) on testing and legal requirements. This benefits both parties—especially the legal aspect. A large corporation like Jetzen probably has an entire legal team employed to handle the various litigation the

Title	Legal Requirements
Source of Requirement	Jetzen Client - IT
Business Requirement	
All content (copy and assets) delivered for publishing must have the proper legal approvals from client All assets must be approved for global use on the web All required product disclaimers will be provided to agency for publishing on the bottom of each page as needed	
Dependencies	
Global legal disclaimers to be provided for publishing: o Copyright and Trademark o Privacy Statement o Ad Choices	
Notes	

Title	Testing
Source of Requirement	Jetzen Client – Marketing and IT
Business Requirement	
This site will need to pass SIT prior to launch This site will need to pass Performance Testing prior to launch Includes Security, Distaster Recover, and Failover	
Acceptance/Fit Criteria	
Agency Quality Assurance team will work with clients to develop test strategies and case studies for use during SIT	
Dependent Requirements	

Figure 3.3 BRDs may include legal and compliance requirements which are often supplied by the client (SIT = System Integration Testing)

company faces. So, let's be clear—as the production company, we're not responsible for any of the content on the website, we just put it there.

Please, Please, Please Read This

The BRD should be shared and reviewed by all parties, but good luck with that. You know people don't like to read in this business! It's one of the most important documents that will be created during the process, and it's very important for everybody to understand what's being built. When I get the feeling that the team isn't reading through the BRD, I sometimes bring it up during a review meeting and go over the details with them. Here's why I feel it's important to make sure that people understand what's in the BRD:

- Some team members may not have a lot of digital experience, so the BRD should be written or presented in layman's terms as much as possible
- The creative team should be looking at the BRD to ensure that their vision has been clearly communicated to, and understood by the tech team
- Finance may want to see the BRD when they go to approve the budget allocation for this project
- The development team will look closely at the functionality descriptions as they create the tools
- The IA team will use the BRD to create their wireframes
- The test team will use the BRD to create their test strategy and cases

Even though we'll be getting this document signed off on before going much further, we don't really lock down scope until *Plan and Define* is over. It is important to solicit input from all parties throughout the Plan and Define period, up until it is time to start development. This is actually one of the hardest parts of our job—how to communicate technical details to nontechnical people. It's especially difficult when we're asking them to read through a large amount of detailed information. I find that most people just don't like to read through this type of material. All I can suggest is to keep hammering home the importance of the information. Convert it into slides—e-mail it out in sections—do whatever needs to be done to ensure that everybody is paying attention.

Once the BRD is done and has been socialized throughout the team, send it to the client for final approval. It's very important to get official acceptance from the client regarding the final version of the BRD, since this document determines what will be delivered (and paid for). Now that we know what the requirements are, I'm sure the client might like to know how much it will cost, and how long the project will take. Ever heard those questions before?

TYPES OF BUDGET ESTIMATES

There are three places within Plan and Define that are appropriate points to provide the client with a budget. Let's review each type of budget and where they fall within our process.

The three types of budgets are:

1. High-Level
2. Preliminary
3. Final

1. High-level Budget Estimate

High-level budget estimates are generally created for two reasons. The first would be when the *client* wants to have a ball-park estimate to figure out if they have enough money to even get started on this venture. This seems to be the scenario portrayed during our phone call with the account person. Another reason would be, if the *agency* is afraid that the client doesn't have enough money to get started on this venture.

As you can see from Figure 3.4, the high-level estimate is provided right at the beginning, before the team even has any real definitions surrounding the potential project. It often seems like an odd predicament when we're asked to provide a budget estimate at this point, some might even call it impossible—but I've figured out a way around it, which I've already explained. But just to hit it again, for high-level estimates, I usually provide three buckets for them; small, medium, and large. The buckets I assign are flexible enough to account for any project that comes around, and are very ambiguous. Just make an educated guess based on some past projects—here's an example of some ranges I might send over.

Small = $50k–$125K
Medium = $125K–500K
Large = $500K+

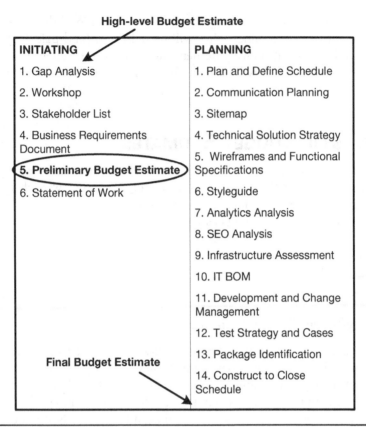

Figure 3.4 This image shows the three typical slots in which a budget may be presented

Seriously, make those numbers whatever you or the sales team want them to be. This is more of a sales tactic than any sort of real budget. Let's review the *pros* and *cons* for providing a high-level estimate:

Pro: On either side of the transaction, client side or agency side, nobody wants to assign resources to a project that has no real chance of getting funded.

Con: The only downside to this scenario is that the clients may derail creativity in hopes of keeping the budget low, instead of creating the optimal customer experience. Let's face it, they either have the money or they don't—but we've all been shopping before and didn't realize what we *needed* until we saw it.

2. *Preliminary Budget Estimate*

This is the route I traditionally take because it's delivered *after* the business requirements are finalized, but *before* Plan and Define starts, so it's a great opportunity to get fairly accurate estimates in front of the client so they can confirm the scope of work. The estimate might fluctuate a bit as we sort out the details, but it should be close unless something unexpected comes up during Plan and Define.

Pro: Having approved scope and budget documents allows the agency to start invoicing for their time and provides focus to the entire team. Regarding invoicing, not too many agencies can start billing without having the client's signature on some sort of purchase order—and the only way to get a purchase order is to prepare a budget. Also, having the approved budget and scope documents means that everything we do during the Plan and Define phase can be measured against them and declared to be either in-scope or added-scope. Again, this provides focus.

Con: At this point in the process, the creative agencies haven't completely thought through their ideas—this will come during the wireframe period. So, when they do try to add scope, it will be the PM's job to deliver the bad news that the *great new idea* will raise the budget or lengthen the schedule. There's always a chance that the client won't want to pay for the increase in scope—then the creative team is angry at the PM for being inflexible and ruining their great new idea.

3. *Final Budget Estimate*

If the team decided to present a preliminary budget estimate, it's still a good idea to confirm the budget at the end of the Plan and Define period, even if there haven't been any changes. Formally label it the *final* budget estimate so the clients understand that these are the final numbers. The steps within Plan and Define have been carefully considered to expose any extra costs to the project, so there shouldn't be any surprises after this point.

Pro: All costs should be captured at this point.

Con: If the client wasn't provided with any budget numbers before this time, they could get a bit of sticker-shock. Also, it's very possible that the agency hasn't been able to submit an invoice for their work up until this time. Finally, I don't recommend attempting to run through Plan and Define without having the preliminary Budget Estimate to use as a stop gate. Those costs provide a line in the sand that the creative team will need to understand or else they'll continue to change the specs of the project without consideration to budget or schedule. If there isn't a budget to refer to during Plan and Define, then the production agency has absolutely no authority to place caution on creativity.

Okay, There's a Fourth Type of Budget

Is it feasible to put together a budget just for Plan and Define, and then another one for the Construct to Close period? Yes, and it's not a bad idea, and in my experience, agencies love this approach—but clients do not. Agencies like it because they're guaranteed a certain payment even if the clients pull the plug after Plan and Define. But the clients don't like it for two reasons:

1. **It sounds too *long*:** Whatever the time period any project is scheduled to last, the client generally wants it more quickly— that's just how clients are. Breaking it up into two projects (one budget for Plan and Define, another budget for Construct to Close) is a psychological obstacle for some people.
2. **It sounds too *expensive*:** The last thing the client needs is to spend money on a Plan and Define, only to find out that they can't afford the Construct to Close. Since there are so many planning activities, the budget for Plan and Define will likely be quite large, and when clients spend that much money, they want something big to hand over to their managers.

CREATING THE BUDGET

When creating a budget pay attention to these two areas which drive costs:

1. Business requirements
2. Resources

To create the budget, first make a list of all of the individual projects listed in the BRD (each requirement). I like to price these out and manage them separately. Having these broken out separately will be a huge benefit during production. First, if one of the projects goes way over budget, we'll be able to track those costs and pinpoint exactly what went wrong and where, then either fix the problem or add it to our lessons learned for next time. Second, if the client needs to decrease the budget they'll be able to easily review what everything costs and simply delete as needed. Third, as the functionality of each project is developed through the course of wireframing, we can easily adjust the individual costs without making the entire budget appear to be ballooning out of nowhere.

So, we're basically making two lists right now. A list of the business requirement *projects* and a list of *resources* that will fulfill not only those projects, but also general work that needs to be done for the site. Let's start with pricing out the BRD tools. Of course, we always ask the resources themselves to provide the number of hours they predict they'll need to spend on each. And remember, we're not just talking developer time for these things—many different people will be working on them:

Business Requirements:
- News Stories RSS feed
- Events calendar
- Video gallery
- Store locator
- Configurator
- Contact us form
- Cascading Style Sheet (CSS)

Possible Resources:
- Front-end developer
- Web developer
- Information architect
- Art director
- Infrastructure manager
- Data architect
- Performance and analytics
- Search optimization
- Project manager
- Asset manager

Doing the Math on Budgets

Start a spreadsheet and ask all the resources how many hours they'll need to bill against each requirement. Then, it's simple math to multiply their hourly rate against the number of hours, and add them up. Table 3.1 shows an example budget for News Stories.

Table 3.1   Budget example for news stories

NEWS STORIES	Hours	Rate	Cost
Information Architect	30	100	$3,000.00
Data Architect	10	105	$1,050.00
Developer	55	85	$4,675.00
Art Director	16	76	$1,216.00
Front-end Developer	16	110	$1,760.00
Project Manager	25	60	$1,500.00
		TOTAL	$13,201.00

Do the same thing for each of the business requirements and just add the final total to the budget estimate that the client will see. The clients really don't need the detail behind the numbers. Whenever I ask the sales team how much pricing detail I should show the clients, they always respond with *as little as possible.*

The other item on our list of cost drivers is resources. What are the resources spending time on if it isn't listed in the BRD? Let's take each of the process steps and list them under each appropriate resource so we understand who's doing what.

Front-end Developer:
- All of these costs should be captured in the requirements, but ask, just in case

Web Developer:
- Most of these costs are captured in the requirements, but there may also be some work needed outside of the BRD, such as the basic framework of the site

Information Architect:
- Gap analysis
- Define business requirements/scope
- Sitemap
- Wireframes
    - ▼ Wires can be priced out as part of the requirements since they will need specifications, but there should also be a set of nontechnical wireframes for the content pages of the site

Art Director:
- Needed to produce a styleguide, but those costs should be part of the CSS requirement and also added to each of the individual requirements or tools

Infrastructure Manager:
- Infrastructure assessment
- IT bill of materials
- 301 redirects
  - ▼ The infrastructure manager doesn't produce the 301 redirect list, but they will be a vital part of implementation
- Cutover management

Data Architect:
- Possibly part of an overall technical solution strategy for the site, and also may be required to bill against individual requirements for those solutions

Performance and Analytics:
- Analytics direction may come from another agency, but we'll need a developer to implement that direction

Search Optimization:
- For Jetzen, search engine optimization (SEO) is coming from another agency and will be applied to the site through our web developer or content author

Project Manager:
- Manage workshop
- Prepare stakeholder list
- Coordinate high-level estimates
- Produce final budget
- Document statement of work
- Communication planning and status reports
- Define business requirements/scope
- Change management
- Package identification
- Construct to Close schedule
- User acceptance cases and testing (UAT)

Asset Manager:
- Content tracker
- Asset quality review

Quality Assurance:
- Test strategy/cases and plans
- Content quality analyses
- System integration cases and testing (SIT)
- Performance testing (security, disaster recovery, failover)

Now that we have all of the tasks broken out, just create a budget for the resources based on the time estimates that they provide. Table 3.2 shows an example budget from the Jetzen project.

The CSS is always one of the biggest numbers and yet, it's a task that nobody ever considers or mentions at the beginning of the project because there's not much to talk about. We need the CSS in order for the website to function, but, not everyone understands what CSS is—it is a basic web development language that integrates the design with html. It transfers the designer's colors, fonts, and layout into html code. This

Table 3.2 Example budget for the entire project

JETZEN BUDGET ESTIMATE

Deliverable	Hours	Rate	Cost
CSS			$80,000.00
Search			$3,500.00
News Stories			$13,201.00
Video Gallery			$17,000.00
Store Locator			$22,000.00
Configurator			$60,500.00
Contact Us Form			$20,000.00
Information Architect	20	100	$20,000.00
Web Developer	40	85	$34,000.00
Infrastructure Manager	3	115	$4,140.00
Analytics Developer	4	95	$4,560.00
Project Manager	30	60	$18,000.00
Asset Manager	59	60	$95,640.00
		TOTAL	$394,541.00

is another reason why we need a styleguide. Art directors who create styleguides specialize in providing the correct specs to front-end web developers so they can transfer the design into CSS code.

Now that we've created the budget, it's *almost* time to show it to the client and get the approval. But, before we do that, it's a good idea to present the budget, along with the scope document (or statement of work) which we'll discuss next.

STATEMENT OF WORK

The statement of work, also referred to as a *scope document* or *project charter*, is a description of the products or services that will be delivered and formally accepted by the client throughout the project, along with how much will be charged for those efforts. For our make-believe Jetzen project, we're not only delivering a website but also the many outputs associated with the 30 process steps, such as:

- Sitemap
- Wireframes
- Contact form
- Bill of materials

All of these outputs will lead up to the final delivery of a finished website. It's important that everybody's on the same page when it comes to what scope is tied to what budget, because even though technical development is detailed and precise, there's also a strong creative component that can escalate and expand if you are not careful to manage it. The following are a couple examples showing the importance of the statement of work:

"That's awesome!" Let's take a look at the configurator requirement. What if, during wireframing, the creative team comes up with a great idea to actually send the customers a miniature prototype of the spaceship that they just configured? I mean, that *is* a super cool idea. The customer would build their spaceship online and then six to eight weeks later a mini spaceship would arrive in their mailbox—looking exactly like the one they built on the website—making them really excited about buying the larger version. Oh, that's a great idea, nobody's going to deny that—and we all want to be a part of it. But trust me, nobody will understand what's so hard about just *e-mailing* the configuration details to a third party who'll build the prototypes and mail them to the customer. Some of the team members will say, "Just e-mail the data out, man. The only thing you might need to do extra is to capture the customer's name and address…"

This is when the PM has to step in and tell the client (and creative team) that they need to measure the extra effort against the original scope and see if there's an increase. Of course there's going to be an increase—probably so large that the client won't be able to afford it, but we'll keep that to ourselves for now. People who speak this sort of truth are ostracized for blocking creative ideas. In front of everyone in the meeting we say phrases like "possible increase" and "let us investigate." I'll talk more about this in the change management section of the book.

"Meh." I can also picture a scenario for News Stories where the client is unhappy with the final deliverable because the articles don't have a *share* feature to easily send them out to social media. Well, I would totally agree that this was an oversight, but it's one we all share. It was never in the BRD or in any of the wireframes. And honestly, it's something that is easy to fix, but we have to measure the cost against the original budget to see what the delta is. Maybe another project is coming up that will be less expensive than expected and we can fit it in, but maybe not. Lots of little increases can balloon into a major deficit if not managed properly.

The statement of work template should reflect not only the project, but also your company and any financial or invoicing service-level agreements that may already be in place for the client. For instance, of course we want to include a list of the deliverables for this project, but maybe we also need to include a payment schedule or type of contract.

Other typical sections in a statement of work include:

Purpose: Why are we doing this project? This is a valid question, and the answer can't be, "Because somebody agreed to pay us for it." Adding a *purposeful* purpose statement provides a clear objective, which can help ascertain the business importance of certain requirements as the project moves forward.

Scope of work: The management team should be consulted regarding the level of detail in this section, but err on the side of listing all deliverables and resources. Do we want to call out that the clients are paying for sixty hours of a data architect and what those hours are to be used for? Or, does the data architect fall under an umbrella of *development* which lists a cumulative price? Another item we can list here, but not in too much detail, is that the website will be responsive. There's no need to say anything other than something like, "ABC Agency will be creating a responsive website for Jetzen Spaceships...."

Assumptions and dependencies: In the Jetzen example, we're noting that hosting is not the responsibility of the agency. This should be a known factor but why not drop it in here just to make sure?

Location of work: This describes where the work is to be performed, for companies having more than one location.

Period of performance: Some companies like to include the office hours their employees will be available to service the clients, or how long the project is scheduled to last. If the client has a lot of internal bureaucracy that gets in the way of progress, then the project could last longer than expected, sucking up valuable resource time and draining profits.

Deliverables or project schedule: This can either be a list of the deliverables and when they're due, or major milestones for the project. It's also nice to provide either a rough draft of the project schedule, or at least a target due date.

Applicable standards: This describes any industry-specific standards that need to be adhered to in fulfilling the contract—perhaps there are page-load times or security requirements to account for.

Acceptance criteria: This section comes in handy if your financial agreement requires the client to officially accept milestones or deliveries before invoicing can proceed. I've been involved with projects where scope has not necessarily been added, but we somehow get into this endless loop of reviews and updates. Make sure there's a clear finish line.

Type of contract/payment schedule: This is important to include if your company has more than one contract in place with a client, or if the payment schedule needs to be made clear.

Figure 3.5 shows an example of a statement of work for the Jetzen project.

The statement of work should be delivered with the BRD and budget estimate, and reviewed and accepted both by the client and a member of the agency's management or finance team. A typical signature area

---

Project Statement of Work – Jetzen Website

1. Revision History

Version	Date	Author	What Changed
1.0	9/28/2014	T. Olson	Initial draft

2. Purpose

To build an engaging website for potential customers of Jetzen Spaceships that promotes product awareness and consideration and preference. Customer arriving to the website may be at various stages of the purchase funnel, and after engaging with the website should be moved downward toward purchase. The objective of this new design is to increase in-store shopping and quotes by 30% for year one.

3. Scope/Project Definition

3.1 The following deliverables will be provided during the process of creating a website:

Stakeholder List
Schedules
Solution Strategy*
Sitemap*
Wireframes*
Mobile Assessment
Style Guide
SIT and UAT Management
Cutover Plan* and Management
*These deliverables will be formally submitted for acceptance

3.2 Please see attached Business Requirements Document for specifics on developmental requirements

News Stories RSS Feed
Events Calendar
Videos Gallery
Store Locator
Configurator
Contact Us Form

Figure 3.5   The purpose of the statement of work is to make sure it's clear what the client is paying for, and what services will be delivered

3.3 Resources
**The following resources will be provided to support this
project. Please see attached Budget Estimate for estimated
hours and rates**
Information Architect
Data Architect
Web Developer
Art Director
Infrastrucuture Manager
Analytics
Project Manager
Asset Manager

3.4 Assumptions and Dependencies
All deliverables will be formally submitted to the client for
acceptance. Once a deliverable has been accepted, any changes
to the deliverable may incur increased costs.
Host provider will be provided by or contracted through Jetzen
Spaceships.
All infrastructure hardware required for this project will be either
contracted through or provided by Jetzen Spaceships.
Agency will provide hosting and infrastructure required for
development and testing prior to cutover.

4. Project Schedule
A formal schedule will be submitted for formal acceptance during
the Plan and Define Period.
We anticipate the total project period to last four to six months.

5. Applicable Standards
In addition to the deliverables listed in section 3.1, the client will be
asked to approve the project prior to cutover.
Agency development environment utilizes Apache secure
webservers with appropriate authentication.
HTTPS security will be used where customer data exchange is
expected.

Figure 3.5 (Continued)

6. **Acceptance Criteria**

   Solution Strategy: Data Architect will work with Jetzen IT to prepare the data strategy with associated polices, capabilities, and systems to meet business requirement needs using enterprise data standards. The strategy will be delivered as a PDF design document. All associated database solutions and infrastructure bill of material requirements are the responsibility of Jetzen Spaceships.

   Sitemap: A sitemap will be submitted for review and approval for up to three rounds of comments. Agency will not move forward with wireframes until the sitemap has been approved.

   Wireframes: Wireframes will be delivered in four sets of deliveries. Each set must be approved before the next set can begin, according to the project schedule.

   System Integration Testing: Each web development tool listed in section 3.2 will be provided online for client review and approval. Agency will coordinate with Jetzen IT for end-to-end testing.

   User Acceptance Testing: An integrated website will be provided online to the client within the agency testing environment for client review and approval. All comments will be captured and monitored using our bug tracking systemand coordinated with creative agency for updates.

   Cutover Plan will be prepared by agency and reviewed with all project team members, including the Jetzen IT and infrastructure team.

   Upon launch, agency will provide sustain support for up to 7 days prior to the official transition to operations presentation which will be prepared by agency to the sustain team.

   All deliverables will be considered accepted via client signature or e-mail confirmation.

7. **Payment Schedule**

   Monthly invoices will be submitted to client and payment expected within 90 days of receipt.

   Web development tools listed in section 3.2 will be invoiced for full amount upon client acceptance.

   All resource time listed in section 3.3 will be billed monthly, not to exceed hourly estimate as stated on budget estimate.

Figure 3.5 (Continued)

is shown in Figure 3.6. Consider these elements your business case for continuing with the project.

Next Steps

With the budget and scope approved by the client, we are ready to move forward with Planning. As a reminder below are the steps for Planning (taken from Figure 1.1 in Chapter 1):

7. Plan and define schedule
8. Communication planning

Client Agreement

Client signature and date below indicates acceptance of this project statement of work, fees and timeline and grants approval to move forward with this initiative based upon project fees and assumptions stated above.

Reviewed and accepted by client:

Signature _____

Name _____

Date _____

PO _____

Reviewed and accepted by agency:

Signature _____

Name _____

Date _____

Figure 3.6 In addition to a signature, the finance team probably requires a purchase order number

9. Sitemap
10. Technical solution strategy
11. Wireframes and functional requirements
12. Styleguide
13. Analytics analysis
14. SEO analysis
15. Infrastructure assessment
16. IT BOM
17. Development and change management
18. Test strategy and cases
19. Package identification
20. Construct to close schedule

These steps do not necessarily need to be completed in this particular order. They're simply a list of different tasks that need to be completed during the planning phase. Throughout the next several chapters of this book, I'll continue to review every step in the process, explaining them from a PM's point of view.

In this chapter we:

✔ Created a BRD
✔ Made people read it
✔ Developed the budget
✔ Wrote the statement of work
✔ Got the client to sign-off on the project

SECTION II: PLANNING

4

COMMUNICATION

I'm always amazed when I notice other project managers (PMs) who don't regularly produce or distribute schedules. It happens more than one might think. What are they using to manage their projects, and how do they get away with it? I like to create the schedule right away and use it as a method for taking control of the process before things start spinning out of control, which could happen early-on since the schedule also serves as our process map.

CREATING THE SCHEDULE

In addition to helping us to manage the tasks of the project, we'll need a schedule right away because the first thing the client will ask right after signing the scope document will be, "When will this thing be ready?" The reason clients ask this is because their managers are asking them the same question and they need to respond back. Their managers don't care about the details of the project—they just want to know how much it will cost, when it will be ready, and then afterward—did it improve sales?

The thing about a digital schedule is that it's a bit fluid throughout the Plan and Define phase, so I usually provide some milestone target dates and then fill in the rest of the details as we move along. Don't get me wrong, we *can* create the entire schedule up front, but it'll probably be wrong shortly after we send it out because at this point in the project, we simply don't have all the information we need to create an accurate schedule. That reminds me of when I used to work on print projects, like the vehicle product catalogs they hand out in dealerships.

They're based on product details that are *constantly* changing, so when the print samples came in we used to gaze at the pictures lovingly, but we'd also say to each other, "Don't read it!" because we were afraid that somebody was going to find a mistake inside.

Digital schedules should be created in chunks, so I'm going to describe the schedule creation process in five steps and explain how they build upon one another:

1. Sitemap
2. Wireframes and styleguide
3. Analysis and development
4. Asset delivery, input, and QA
5. SIT, UAT, and cutover

But first, let's look at a basic engagement model that most agencies use. Once I explain this process I won't have to explain every single section of the project plan because, for the most part the reviews are all handled the same way.

The important thing to note about Figure 4.1 is that all deliverables are reviewed internally before being shown to a client. That way the agency is showing the clients their best work that's already been vetted out, exposed to scrutiny, and updated appropriately. All of the deliverables in our schedule will be handled the same way. At the end of the process the deliverable is sent to the client for approval. Clients provide their approval two different ways and I want to make sure it's understood that the approval needs to be captured. If the client provides verbal approval during a meeting, then the PM should document that approval in the meeting notes. If the client sends an e-mail with their approval, then I like to create a PDF of the e-mail and save it to the project files for future reference. Let's be honest—this is how we cover ourselves if something goes wrong.

Step One—Sitemap

When creating a schedule I like to include the dates that the gap analysis, business requirements document (BRD), and statement of work were finalized, just to capture a complete project history. Next, we need to get the team prepared to tackle the sitemap, which will provide direction for the rest of the project. I've started a basic project plan for this deliverable (see Table 4.1), but please note that the number of days

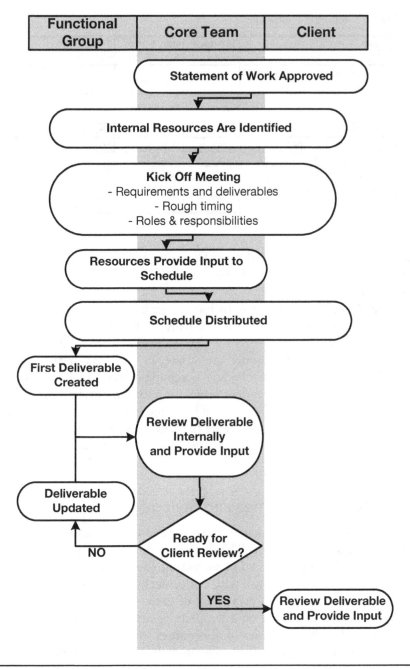

Figure 4.1   All deliverables are reviewed internally prior to the client review so potential errors can be fixed beforehand

Table 4.1 Top of the schedule

1	**Jetzen**	**152 days**	**Mon 1/5/15**	**Thu 2/12/15**		
2	Gap Analysis Presentation	0 days	Mon 1/5/15	Mon 1/5/15		IA
3	Business Requirements Finalized	0 days	Mon 1/26/15	Mon 1/26/15		IA,PM
4	Statement of Work Approved	0 days	Fri 1/30/15	Fri 1/30/15		
5	**Sitemap**	**9 days**	**Mon 2/2/15**	**Thu 2/12/15**		
6	Sitemap Developed by IA	5 days	Mon 2/2/15	Fri 2/6/15		IA
7	Team Review of Sitemap	1 day	Mon 2/9/15	Mon 2/9/15	6	ALL
8	IA Incorporates Changes	2 days	Tue 2/10/15	Wed 2/11/15	7	IA
9	Final Sitemap Review and Approval of Sitemap	1 day	Thu 2/12/15	Thu 2/12/15	8	Client

assigned to each task is arbitrary—for the purpose of this book, and should be provided by the resources that will actually be creating the deliverables.

This should be pretty easy to follow. An information architect (IA) creates the sitemap; we review it internally; then if there are any changes, the IA can make them before it's sent to the client. I'll dig into the specifics of the actual sitemap document later in this book. In this chapter we're just focusing on the schedule—creating it and explaining how the different sections are dependent upon one another. But, let's take a look at the sitemap we created for Jetzen (see Figure 4.2), because we'll need to look at that before we can create the schedule for the wireframes and styleguide. We're going to use it to decide on what our sprint groups should be.

Step Two—Wireframes and Styleguide

Now that we have a sitemap, we can start to take an agile approach to the production assignments. The dictionary describes the adjective *agile* as *quick and well-coordinated in movement,* and this certainly applies to what we want to accomplish with our project plan. We'll need to create a wireframe for just about every page of the website, with the exception of pages which are based on a template or are repeatable. But, instead of waiting for the entire set of wireframes to be ready before we can start the review process, we'll take an agile approach by breaking the sitemap into groups that the IA can handle one at a time. This way, after each group of wireframes is ready, we can move them on to the next phase of production (styling), while the IA starts on the next group of

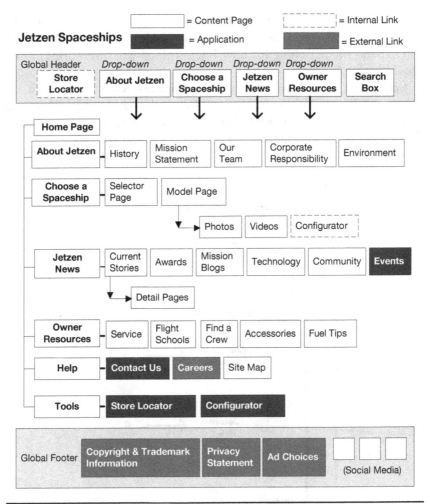

Figure 4.2 Sitemaps organize the content and provide an overview of the experience

wireframes, and so forth. This way we keep several resources working at the same time in a *quick and well-coordinated* manner.

To better understand this concept, let's take a look at the sitemap and decide what the various groups should be. The IA can help you with this, since they'll have a better understanding of how much work will be involved with each section. By the way, these *groups* can also be called *sprints*, *bundles*, or *packages*—take your pick. For this book we're going to call them sprints.

I divided the work in the way that I thought it would actually be done in the real world. So take a look at Figure 4.3, and follow along as I explain my reasoning for how I determined what the sprint groups would be for Jetzen.

Sprint 1: Global Navigation and Homepage—Always start with the framework. The rest of the website will follow the basic width and column rules set up with this set of wireframes. Plus, the creative agency normally designs and gets approval from the clients on the homepage first.

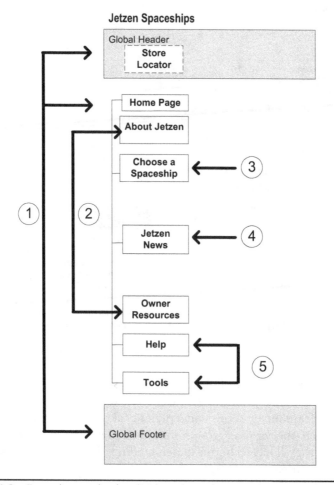

Figure 4.3 Formulate a plan by dividing the sitemap into manageable work packages, or "sprints"

Sprint 2: About Jetzen and Owner Resources—When reading the page titles on the sitemap it seems like these pages will be fairly easy. There's not a lot of development needed for these pages—just good content. That being the case, the creative agency will probably tackle those pages next.

Sprint 3: Choosing a Spaceship—This is a very important focal point of the website, since these pages detail the actual product. They contain several *tricky* pages; the model main page will be a virtual catalog of each spaceship model and will need to *work hard* to sell spaceships. Associated with the main page will be the photo and video pages which have to be included with this group as well. This group will be more difficult than Sprint 2.

Sprint 4: Jetzen News—These pages may be a bit difficult too, as the media feed and filter tools need to be thought through and mapped out. We could make a strong argument to tackle the hardest pages first in an effort to get them into development more quickly, since they'll probably take a long time to create—but from my experience, creative teams like to start on the easier pages first. (They wouldn't be the only ones to ever procrastinate; ask me how long it took to write this book.)

Sprint 5: Help and Tools—This group is also tricky, because three application pages are part of this group: the contact form, the store locator, and the configurator.

So why was it important to wait until the sitemap was approved before we started on the wireframes? One reason is so that we had a solid production plan associated with the sprint groups we put together. The sitemap provides a few other critical details we'll need moving forward:

- When developing the sitemap, the IA imagines how customers will experience the site as they click from page to page. The pages aren't just thrown together randomly; they're sequenced—like pages of a book—to tell a story. It's nice to know which pages work and connect together in order to create the wireframes.
- The global header and footer are very important components of the homepage, which is always brought into Sprint 1. Make sure everybody is happy with the way the sitemap is laid out, because if not, a lot of associated deliverables will have to change later on.
- Application pages are identified, although not completely designed at this point, and because they require hardcore development time, they'll be more work to create than *content* pages. It's important to balance out the sprints in terms of workload, so the team can work efficiently.

Now that we have our sprints approved, we can build out the next part of the schedule (see Table 4.2).

The agile approach comes into play as the IA finishes up the Sprint 1 wireframes and then starts on the Sprint 2 wireframes, while the designers start styling Group 1, and so forth. Each of our sprint groups follows the exact same process—as you can see from looking at Groups 1 and 2 in Table 4.2. It also shows that there are four resource names we're working with in this section: IA, client, design, and creative. In order to ensure that everybody feels comfortable with the schedule, it's a good idea to speak to each of them—separately or as a group—to explain the sprints, and also to ask them to determine how many days will be needed for each of their tasks. Keep in mind, if we have more than one IA available to work on this project—or more than one design team— then that will make a difference when it comes to sequence and timing. For the Jetzen project, we're assuming that we only have one of each resource type.

Also from this example, note that as soon as the first set of wires are approved on Line 14, the IA starts working on Sprint 2 wireframes.

Table 4.2   Part two of the schedule

10	**Group 1 - Navigation and Homepage**	**25 days**	**Fri 2/13/15**		
11	Wireframe Development	5 days	Fri 2/13/15	9	IA
12	Wireframe Presentation	1 day	Fri 2/20/15	11	IA
13	Revisions Distributed to Team	2 days	Mon 2/23/15	12	IA
14	Client Approval	1 day	Wed 2/25/15	13	Client
15	Comp Development	10 days	Thu 2/26/15	14	Design
16	Styleguide Development	5 days	Thu 3/12/15	15	Creative
17	Styleguide Presentation	1 day	Thu 3/19/15	16	Creative
18	**Group 2 - About Jetzen and Owner Resources**	**26 days**	**Thu 2/26/15**		
19	Wireframe Development	5 days	Thu 2/26/15	14	IA
20	Wireframe Presentation	1 day	Thu 3/5/15	19	IA
21	Revisions Distributed to Team	2 days	Fri 3/6/15	20	IA
22	Client Approval	1 day	Tue 3/10/15	21	Client
23	Comp Development	10 days	Thu 3/12/15	15	Design
24	Styleguide Development	5 days	Thu 3/26/15	23	Creative
25	Styleguide Presentation	1 day	Thu 4/2/15	24	Creative

Follow this example for the rest of the tasks (see Table 4.3). If we only have one team approaching the work they can probably only do one group or task at a time. Now to complete the schedule just keep cascading the work down until all of the groups are complete.

The same methodology applies to the comp development from the design team and styleguide development from creative. There may be some downtime in between tasks for the resources, but try to keep everything moving. Keep in mind also, that the resources may have other projects they're assigned to, so always make sure they're comfortable with the dates before publishing them to the extended team.

An experienced PM will notice that there are some dates missing from my timeline. I have *comp development* as one task, when in reality there are many sub-tasks associated with getting comps approved. When a deliverable is due from another agency, some PMs don't list all of the other agency's tasks because that information isn't absolutely

Table 4.3 Agile movement of the IA

Group 3 - Choosing a Spaceship		
Wireframe Development		IA
Wireframe Presentation		IA
Revisions Distributed to Team		IA
Client Approval		Client
Comp Development		
Styleguide Development		Creative
Styleguide Presentation		Creative
Group 4 - Jetzen News		
Wireframe Development		IA
Wireframe Presentation		IA
Revisions Distributed to Team		IA
Client Approval		Client
Comp Development		Design
Styleguide Development		Creative
Styleguide Presentation		Creative
Group 5 - Help and Tools		
Wireframe Development		IA
Wireframe Presentation		IA
Revisions Distributed to Team		IA
Client Approval		Client
Comp Development		Design
Styleguide Development		Creative
Styleguide Presentation		Creative
Scope Freeze		

necessary. As long as the PM knows when to expect the deliverable—
that's all the information they want. Other PMs like to list every single
line item, and the clients might appreciate this as well since they have to
keep everybody on track. Table 4.4 shows what a typical comp development process might look like.

Now that every aspect of the website has been wireframed, designed,
and styled, it's time to call *scope freeze* (see Line 60 of Table 4.5). Scope
freeze is such a wonderful time of year, isn't it? This means that any
functional or design changes beyond this point will be reviewed against
the current budget and schedule, and may result in revisions to one or
both of those documents. This will be described in further detail in the
chapter on development, but just understand that if we don't have a
scope freeze period, both the clients and the creative team will keep
updating the requirements—trust me. And constant change makes a
project unmanageable.

Table 4.4   Schedule for comp development

31	**Comp Development**	10 days	Thu 3/26/15		
32	Initial Creative Development	3 days	Thu 3/26/15	23	Design
33	Internal Review	1 day	Tue 3/31/15	32	
34	Comps Are Updated	0 days	Wed 4/1/15	33	Design
35	Client Review	1 day	Wed 4/1/15	34	Client
36	Copy Development	1 day	Thu 4/2/15	35	Copy
37	Comps Are Updated	1 day	Fri 4/3/15	36	Design
38	Proofreading Review	1 day	Mon 4/6/15	37	Proof
39	Final Client Review	1 day	Tue 4/7/15	38	Client
40	Legal Review	1 day	Wed 4/8/15	39	Legal
41	Delivery	0 days	Thu 4/9/15	40	Design

Table 4.5   Schedule for technical prep

59	Styleguide Presentation	1 day	Thu 5/14/15	58	Creative
60	**Scope Freeze**	**1 day**	**Fri 5/15/15**	**59**	
61	**Analysis and Final Prep**	**13 days**	**Fri 5/15/15**		
62	Create Test Strategy/Cases	10 days	Fri 5/15/15	59	QA
63	Review Test Strategy/Cases	3 days	Fri 5/29/15	62	PM
64	Analytics Analysis	10 days	Fri 5/15/15	59	Analytics
65	SEO Analysis	10 days	Fri 5/15/15	59	SEO
66	**Development**	**82 days**	**Fri 3/20/15**		

Step Three—Analysis and Development

Depending on the project, there can be any number of tasks that need to be complete before development starts. In the case of the Jetzen project, I have dropped a few line items in this group just to make sure we don't forget to get them done—including testing documentation, and search and analytics analysis. If we needed to set up development environments for this project, I would certainly put that in this area as well. Analysis and Final Prep (see Table 4.5) often contain tasks that don't fall into the critical path and can be completed in tandem with other assignments.

So with that, why did we need to wait for wireframes to be in the schedule before dropping in the dates for the Analysis and Final Prep tasks? Well, for one thing, there are two levels of service for both search engine optimization (SEO) and analytics. On the first level we have client-facing people who should be part of the team from the beginning, not only to understand the client's business requirements and the strategy behind the design, but also to provide recommendations on how to optimize the site. On the second level there are developers who need to ensure that the code is in place to deliver on those desired outcomes. That's the part that goes into the analysis area. Once we have a complete set of wireframes and scope has been frozen, it's a good time for those analytics and SEO developers to look at the requirements and make sure they can deliver on them.

Development can't (or really shouldn't) start until the wireframes and styleguide are complete as they have a *finish-to-start* precedence relationship. This is because the wireframes and styleguide fully describe the expected functionality, and with this knowledge several other tasks can begin—such as development, test strategy, and test cases. Also, in terms of the schedule, we can't ask the developers to give us their estimated number of days for development until they fully understand what they need to do. Inputs to developing the test strategy and cases include:

1. BRD
2. Sitemap
3. Wireframes
4. Styleguide

The quality assurance (QA) team is responsible for test strategy and cases, and they can be completed in tandem with other tasks. I don't ask the client to review the test documents because, honestly, they shouldn't have to. The internal team understands what the requirements are and should ensure, on the client's behalf, that the testing documentation will support the test team's evaluation of whether or not the developers are meeting these standards.

Now it's time to talk about development (see Table 4.6). I'm not going to go into too much detail about this subject right now, because there are so many different ways to approach coding and testing, and your company may have a specific way of doing things. Since this chapter is about putting together a preliminary, fluid, and flexible schedule, that's all we're going to do right now. We'll have a kick-off meeting with the development team, show them all of the documentation (wireframes will be most important), ask for their time estimates, and put those estimates into the schedule so the team knows approximately when we will be starting systems integration and user acceptance testing.

As you can see from my predecessor line numbers on the right of the chart in Table 4.6 (17, 51, etc.), I have the development starting at different time intervals based on when the styleguide for each of those sections will be complete. For instance, New Stories is part of Sprint 4 and styling for that sprint will be complete on April 30, so that development begins on May 1. Later we'll add the testing dates to the schedule, but for now, just make a list of each of the development tasks and input their time estimates.

Step Four—Asset Delivery, Input, and QA

The trick to asset delivery, input, and QA is to try and coordinate it so that the end of this process coincides with the end of development. While it may seem proactive to get the assets delivered and input as soon as possible, doing it too early may result in the assets being out-of-date

Table 4.6   Schedule for development

66	Development	82 days	Fri 3/20/15	Mon 7/13/15	
67	CSS	50 days	Fri 3/20/15	Thu 5/28/15	17
68	HTML Framework	60 days	Fri 3/20/15	Thu 6/11/15	17
69	Search	4 days	Fri 3/20/15	Wed 3/25/15	17
70	News Stories	25 days	Fri 5/1/15	Thu 6/4/15	51
71	Video Gallery	14 days	Fri 4/17/15	Wed 5/6/15	43
72	Configurator	42 days	Fri 5/15/15	Mon 7/13/15	59
73	Contact Us Form	12 days	Fri 5/15/15	Mon 6/1/15	59

by the time the entire deliverable is ready to launch. For Jetzen it looks like development will be wrapping up on July 13—from Line 72 of the chart in Table 4.6.

Like any part of the schedule, the PM should run the dates by the pertinent team members to make sure they're comfortable with their assignments and timing. For this part of the schedule (see Table 4.7), I usually take a whack at it myself and then give the team some time to review and request changes. Starting with *asset delivery*, I would ask the creative team if they want to deliver the assets according to the same sprint groups we came up with during the wireframing, but if not, to simply provide me with a list with the specific pages that will be delivered with each group—make sure they don't miss any pages. For our example project, let's say the creative team will deliver the assets in three deliveries—June 1, 8, and 22nd.

For *Asset Quality Review* (AQR) I give the asset manager a couple of days per group to review the assets to make sure everything was delivered according to specification. For instance, do we have every image we were expecting and every piece of copy? Are the images the correct size and weight? Part of our risk mitigation plan for this project will be to schedule a couple of days for the creative team to make *AQR Fixes*

Table 4.7 Schedule for the content tasks

74	Asset Delivery	16 days	Mon 6/1/15		
75	Group 1	1 day	Mon 6/1/15		Creative
76	Group 2	1 day	Mon 6/8/15		Creative
77	Group 3	1 day	Mon 6/22/15		Creative
78	Internal AQR	17 days	Tue 6/2/15		
79	Group 1	2 days	Tue 6/2/15	75	Asset Mgr
80	Group 2	2 days	Tue 6/9/15	76	Asset Mgr
81	Group 3	2 days	Tue 6/23/15	77	Asset Mgr
82	AQR Fixes	16 days	Thu 6/4/15		
83	Group 1	1 day	Thu 6/4/15	79	Creative
84	Group 2	1 day	Thu 6/11/15	80	Creative
85	Group 3	1 day	Thu 6/25/15	81	Creative
86	Content Input	20 days	Fri 6/5/15		
87	Group 1	5 days	Fri 6/5/15	83	Dev 1
88	Group 2	5 days	Fri 6/12/15	84	Dev 2
89	Group 3	5 days	Fri 6/26/15	85	Dev 3
90	Internal QA	17 days	Fri 6/12/15		
91	Group 1	2 days	Fri 6/12/15	87	QA
92	Group 2	2 days	Fri 6/19/15	88	QA
93	Group 3	2 days	Fri 7/3/15	89	QA

to the assets that did not pass AQR. It's very typical to find errors that need to be fixed, or undelivered assets, so don't stress if this happens—that's why we're adding the AQR Fix dates.

Once the assets pass AQR, they should be sent to whomever needs to insert them (content input), whether that resource be a developer or content author. This part will take the longest, so plan on about five days per group, if you're just guessing—until you can run it past the appropriate team members for a more accurate count. Finally, once the assets are input and the code is ready we move into *internal* QA. Again, notice the agile approach we are taking with the asset process. Follow each group down as it moves through the process—keep it moving!

Step Five—SIT, UAT, and Cutover

Systems integration testing (SIT) and user acceptance testing (UAT) can be done in a number of ways and it differs not only between companies, but also between projects done at the same company. Before we get started, if the testing schedule seems a bit aggressive, that's because the development has been completed and tested in a lower environment before entering the preproduction environment. We fully test any code prior to letting the extended team conduct their testing to make sure it's working properly. It's really aggravating when somebody asks you to test something that obviously doesn't work. It's for sure expected to work once the clients and extended team members do their tests. They should just be validating the tools at this point—not having to log errors. Table 4.8 is an example of a waterfall approach where we wait until all of the development is complete before testing begins. There are a couple of possible benefits to this approach, if you choose to use it:

1. Code fixes can be all done at the same time and deployed together, which saves on internal resource time.
2. Even though the development is staggered, each of these individual projects will be launched at the same time so it doesn't really matter if some of the development gets done sooner.

However, we're not using a waterfall approach for this project. We've already said that we're using agile. See Table 4.9 for an example of the same schedule using an agile approach. Notice that we begin SIT testing as soon as each tool's development is complete in the lower environment. The benefit to getting the testing done as soon as the tools

Table 4.8   Schedule for waterfall-style testing

SIT	11 days	Tue 7/14/15	Tue 7/28/15
Code Deployment to Pre-pro	1 day	Tue 7/14/15	Tue 7/14/15
Internal Review	2 days	Wed 7/15/15	Thu 7/16/15
Code Fixes	2 days	Fri 7/17/15	Mon 7/20/15
End-to-End Testing	3 days	Tue 7/21/15	Thu 7/23/15
Code Fixes	3 days	Fri 7/24/15	Tue 7/28/15
UAT	**10 days**	**Tue 7/21/15**	**Mon 8/3/15**
Creative Review	5 days	Tue 7/21/15	Mon 7/27/15
Fixes	3 days	Fri 7/24/15	Tue 7/28/15
Final Team Review	3 days	Wed 7/29/15	Fri 7/31/15
Fixes	1 day	Mon 8/3/15	Mon 8/3/15
Cutover	**1 day**	**Tue 8/4/15**	**Tue 8/4/15**

Table 4.9   Schedule for agile-type testing

SIT	55 days	Thu 5/7/15	Wed 7/22/15
Video Gallery	**8 days**	**Thu 5/7/15**	**Mon 5/18/15**
Code Deploy to Pre-prod	1 day	Thu 5/7/15	Thu 5/7/15
Internal Review	2 days	Fri 5/8/15	Mon 5/11/15
Code Fixes	2 days	Tue 5/12/15	Wed 5/13/15
Update Release Package Code	1 day	Thu 5/14/15	Thu 5/14/15
End-to-End Testing	1 day	Fri 5/15/15	Fri 5/15/15
Final Review and Fixes	1 day	Mon 5/18/15	Mon 5/18/15
Contact Us Form	**7 days**	**Tue 6/2/15**	**Wed 6/10/15**
Code Deployment to Pre-pro	1 day	Tue 6/2/15	Tue 6/2/15
Internal Review	1 day	Wed 6/3/15	Wed 6/3/15
Code Fixes	1 day	Thu 6/4/15	Thu 6/4/15
Update Release Package Code	0 days	Thu 6/4/15	Thu 6/4/15
End-to-End Testing	2 days	Thu 6/4/15	Fri 6/5/15
Final Review and Fixes	3 days	Mon 6/8/15	Wed 6/10/15
News Stories	**9 days**	**Fri 6/5/15**	**Wed 6/17/15**
Code Deployment to Pre-pro	1 day	Fri 6/5/15	Fri 6/5/15
Internal Review	1 day	Mon 6/8/15	Mon 6/8/15
Code Fixes	2 days	Tue 6/9/15	Wed 6/10/15
Update Release Package Code	1 day	Thu 6/11/15	Thu 6/11/15
End-to-End Testing	1 day	Fri 6/12/15	Fri 6/12/15
Final Review and Fixes	3 days	Mon 6/15/15	Wed 6/17/15
Configurator	**7 days**	**Tue 7/14/15**	**Wed 7/22/15**
Code Deployment to Pre-pro	1 day	Tue 7/14/15	Tue 7/14/15
Internal Review	1 day	Tue 7/14/15	Tue 7/14/15
Code Fixes	1 day	Wed 7/15/15	Wed 7/15/15
Update Release Package Code	1 day	Thu 7/16/15	Thu 7/16/15
End-to-End Testing	1 day	Fri 7/17/15	Fri 7/17/15
Final Review and Fixes	3 days	Mon 7/20/15	Wed 7/22/15

become available is that it leaves room just in case one of the tools is way off and not working properly. This way we can get the easy ones out of the way and concentrate on the tools that need a little more TLC.

So that's the schedule. A complete project plan all in one piece (pretty much) can be seen in Appendix A of this book. The process that I'm describing in this book works for every type of digital project. Sure, we're not always dealing with a sitemap, but there will be some sort of solution map or data flow process to follow. Think about breaking apart an engine, figuring out how it works, and then putting it back together again. Like an engineer we must work with the team to decipher the best approach to rebuilding the engine—which pieces need to go on first and what they are connected to. Then we lay out the plan and follow it.

> The *Project Management Body of Knowledge* describes critical path as, "… the longest path through the project."—that is, the tasks that will take the longest time to complete sequentially and also, are critical to complete. So, just for fun, can you determine what the critical path would be for Jetzen? Look for the answer in Chapter 8 describing our Construct to Close activities.

Please remember that we're not expected to come up with this schedule all at once in the beginning of the planning phase. In fact, the list of 30 rollout steps shows that the schedule isn't completed until the last step in planning. There are going to be people asking over and over again for a launch date, but we have to be very cautious when making a prediction. At this point in my career, I can usually give a pretty accurate quote, but I always tell them that the final date won't be determined until the end of the planning stage. Before that time we can only make a rough estimate. Here's the cadence:

1. Complete the schedule up to sitemap approval.
2. Agree on what the sprint groups will be and how long they'll take.
3. Complete the schedule through wireframes and styleguide.
4. Once the wireframes and styleguide are actually complete (the work has been done) the developers can look at the functionality and provide their timing.
5. Complete the schedule through analysis, final prep, and development.

6. Coordinate the asset delivery and input to be done at the same time as development, followed by SIT, UAT, and cutover.

Piece of cake, right? We're really moving now!

COMMUNICATION PLANNING

In the life of a PM there are few areas as important as project communication. It can be argued that a PM shouldn't be expected to know much about development or end-to-end testing, but communication, schedule, and budget are all ours for sure. It's our job to monitor and control the project, manage stakeholder expectations, and communicate out to the entire group. No disagreement there, right?

INTERNAL KICK-OFF MEETING

Once we have Initiation under our belts and have put together at least the first steps of the project schedule, we should call the internal team together and get everybody on the same page. During this meeting we request representation from all internal departments and resources. Start from scratch and go through the same things covered during the original workshop, plus any new documentation (such as the schedule) that's available.

Internal Kick-off Agenda

- Project name and objectives
- Business lead (client)
- Agency partner roles and responsibilities
- Review of comps or initial requirements
- Gap analysis presentation
- Schedule

The objective of this meeting is to ensure that everybody understands what they'll be responsible for, especially for the very next steps. For the Jetzen project, we should be reviewing the sitemap in about five days, according to the schedule as shown in Table 4.10. Most important,

Table 4.10   Sitemap schedule

5	Sitemap	9 days	Mon 2/2/15	Thu 2/12/15		
6	Sitemap Developed	5 days	Mon 2/2/15	Fri 2/6/15		IA
7	Team Review of Sitemap	1 day	Mon 2/9/15	Mon 2/9/15	6	ALL
8	IA Incorporates Changes	2 days	Tue 2/10/15	Wed 2/11/15	7	IA
9	Final Sitemap Review and Approval of Sitemap	1 day	Thu 2/12/15	Thu 2/12/15	8	Client

confirm with the IA that they'll have a sitemap to review on February 9, and then open up the floor to questions before closing the meeting.

Status Reports

Once we have the sitemap ready for client review, we should also have a status report for the team to follow along with. The status report should list every single task that we'll be working on, what the status is, and any risks and mitigations associated with the task. I've always said that nothing brings a project to life like a status report. So many times I've been briefed on a project that seemed overwhelming; the requirements were undefined and there was no plan in place. Nonetheless, once I put together a status report to follow, the team rallied around it and we started chipping away. But, there are so many different types of status reports—which one is best? It needs to fit not only the complexity of the project, but also the personality of the PM.

As with all of the documentation needed to manage a project, the PM should adjust the templates to what suits them as owners of the process and communication. When I train new PMs, I give them examples of what I use, but at the same time, urge them to alter the templates as necessary into what they're comfortable using. Some PMs prefer reports with a large amount of detail, listing all of the risks, mitigations, and key dates, while others are more comfortable with less information. The key is to clearly communicate what all of the tasks are that need to be completed, what their sequence is, when they're due, who they're assigned to, what the status is, and what, if any, roadblocks there are. That part about naming the owner is really important. I have definitely been part of projects where the owner was never specified, so nobody ever did anything. We can all look around the room agreeing that the task needs to be done, but unless we tell Joe it's his job, nobody's going to do it—call out specific names.

Is the status report starting to sound a lot like the schedule? Well, it is—except for the actual *status* part, including the commentary needed to explain each step's description and issues. Let's review a few different types of status reports:

1. Table Status Report

This is a very typical type of status report, which we've all seen before—it's usually created using tables in *Microsoft Word*. Most agencies are comfortable with this type of report, and it can be used for both simple and complex projects. What I like about the table status report is that it clearly lists every task name, and then has a space for status information, next steps or issues, and of course the task's owner name (see Table 4.11). It also provides a vertical sequence of deliverables. Another great feature of the table status report is that it can go on forever and include an enormous amount of detail. (I once remember making a twelve page status report for a huge project and feeling so in love with it once I was done; that's the odd nature of a PM for you!)

2. Online Status Reports

Many online open source status report and schedule tools are available and may be useful to the team. For instance, there are kinds that merge the schedule with the status report so that the team only need review and update one document. That's quite helpful! Also, for dispersed or virtual teams that need to collaborate on documentation these tools are perfect for sharing and updating remotely. Personally, I'm not too keen on the *sharing* approach, because I like to update the documents myself. I find that if other people are updating their own status and schedules, I don't pay attention as closely as I should, considering I'm the lead PM. On top of that, I like things a *certain way*, if you know what I mean—don't be messing up my formatting or alignment. In my opinion the online status report should only be used for simple projects because they're not really convenient for listing a large amount of detail. To each his or her own, I guess.

3. Vertical Status Report

Vertical status reports (see Figure 4.4) are often used for simple projects that require just a quick update with few details. The benefit to this

Table 4.11   Example of a table status report

TASK	STATUS	NEXT STEPS / ISSUES	LEAD
SITEMAP	• Complete	• Final version distributed to team February 13	B. Roberts
WIREFRAMES	• Navigation and Homepage Complete • About and Owners Complete • Choosing a Spaceship currently being revised by agency per client comments	• Choosing a Spaceship revisions will be delivered to client March 20 • News Stories scheduled to begin development March 24 • News Stories presentation March 31	G. Lee
COMPS	• Navigation and Homepage delivered, but working on update to masthead treatment • About and Owners sections under development	• About and Owners sections scheduled to be delivered March 25, although legal review has been delayed. • Due to delay, creative team has started advanced work on Group 3 comps	P. Rich
STYLEGUIDE	• Navigation and Homepage presentation scheduled for March 19 @ 3pm • This review is most important for creative teams	• Agency requests update to Homepage masthead treatment with the About and Owners delivery so we can include it with the second styleguide presentation	K. Layman
TESTING DOCUMENTS	• Development of Test Strategy and Cases documentation will begin upon receipt of final styleguide	• Agency will review test documentation with clients week of June 1	S. Kelly

version is that the PM can quickly convert the document into notes to be distributed after the meeting. It's really an agenda, status report, and meeting notes all in one. There's nothing wrong with this procedure at all—I use it quite often.

There are many other types of status reports. Heck, we can create our own never-seen-before report templates—it doesn't really matter, as long as the report serves its purpose of monitoring and communicating.

Status Report

Meeting Purpose: **Jetzen Spaceship Website Status**

Date: **June 11, 2015**

<u>Project Status</u>
- Developmemt
  - o   Domain and v-host have been secured, implemented and tested
  - o   Currently building in development environment
- Content Status
  - o   Navigation and Homepage – **QA Passed**
  - o   About and Owners – **Currently in QA**
  - o   Choosing a Spaceship – **Content Entry Period**
  - o   News – **Going through AQR**
  - o   Help & Tools – **Expected June 22**
- Analytics
  - o   Report Suite was set up
  - o   Input js file then send out link for testing
- SEO
  - o   SEO team sent copy recommendation – **Input in Progress**
  - o   SEO team to provide SEO recommendation for images – **Due June 25th**

Figure 4.4   Vertical status reports are easy to review, update, and distribute

Some people like to use elaborate reports with a section that includes red, yellow, and green status indicators, followed by a list of all risks and mitigations, and then a summary of what was completed last week, what's being done this week, and what will be due next week—they can go on and on. The template we use should be according to our personal preference, but remember that nobody wants to sit in a status meeting all day, so please design it to be concise and let the team get back to work.

Other Types of Documentation

There are a few more documents that may be required for any given project. None of them are necessary to create for every single project, but they might be helpful. Don't create something just to look like a superior PM who knows all of the different types of documentation—everything we do should be to the benefit of the project and team members.

Risk register: Risks are problems that are not necessarily anticipated, but noted as things that could potentially happen that would either slow down or completely derail the project. *Risk registers* log each of the risks along with their *mitigations*, which state how the risks will either be avoided or how the resulting problem will be fixed, if in fact the risk does occur. I think people like to document risks because it makes them feel covered in case something happens—as if we would blame them for it happening. Most risks are obvious, and it feels odd to write them down as something that might possibly happen, but if they want a risk register just start marking them down along with their mitigations or impacts.

Issues log: In addition to listing issues in the status report, another option is to create an *issues log* which lists all issues that need to be addressed and closed. Creating a spreadsheet for issues allows them to be identified by owner, as well as by type or severity. Open, in progress, and closed issues are logged and tracked throughout the duration of the project. This system is helpful not only for tracking issues on the current project, but detailed notations regarding the root cause analysis and problem solving tactics help future projects avoid these risks.

Executive summary: Upper echelon management or stakeholders who are not part of the core project team may require a summarized version of the status report that provides just enough information to comprehend the position of the project. There's really an art to creating these summaries as they require a mind that understands how executives think, what they're concerned about, and how the project might affect decisions they make on any given day. Summaries should not be longer than one page and should clearly communicate the highlights of the project using bullet points, charts, and other visual graphics.

Meeting notes: Now that a status report meeting has been conducted, it's expected that the PM was capturing notes during the discussions and that meeting notes will be distributed to all team members, regardless of whether or not they attended the meeting. Key areas to document include the subject of the meeting, the date on which the meeting was held, who attended the meeting, what was discussed or decided during the meeting, and what the next steps are.

If I had to list one trick to the art of note-taking at meetings, it would be to capture not the entire discussion, but just a summary of what the issues were, and what was decided. For instance, during the Jetzen status meeting, perhaps there was a lengthy discussion about what sort of building tools should be included with the spaceship configurator. Should we include the engines, types of air pumps, available seats, option prices, etc.? No need to list all the, "he said, she said," back and forth—just mark down that the discussion took place and what the final outcome was.

I do find it very helpful to pull out *next steps* or *action items* from the meeting notes and list them in a summary at the bottom, because most people who actually attended the meeting won't read through the notes. They were there. They lived it. And yet, since they won't read the notes, it's quite possible that they'll forget what they were assigned to follow up on and when it's due. A lot of people just look for their name at the bottom of the report, while thinking "Am I supposed to be doing something?"

Version Control

Another form of project communication is version control. When sending out updates to documents that have previously been distributed, it is crucial to alter the version nomenclature on the naming convention of the file. For instance: *Jetzen_Schedule_v2*

Dates can also be used as the version number, but I don't really recommend this approach as we may find ourselves sending out multiple versions on the same day. Additionally, "v2" and "v3" are much easier to read than dates like *100215*. I always have to study those—and forget about it if you're on a global team—they list dates differently than we do.

One more note on sending out documentation; it's good practice to check where the task is in real-time versus where it is supposed to be according to the schedule, and append the e-mail with an indication of what's due and when. For example, if the Jetzen wireframes for Sprint 1 are distributed two days behind schedule, then the result will probably be that the subsequent tasks will be delayed. Try to determine how the project can get back on track. Can the next task start with a partial delivery? Or, can they trim down the number of resource hours needed to complete the task? Perhaps the review period for the wireframes can be shortened?

Finally, remember not to take bad news personally. The PM rarely causes delays or problems, but we're always responsible for communicating those problems out and tracking down the solution or mitigation. Some PMs hide bad news from the team because they fear the team will blame the messenger. Well, sometimes they do. People get bad news and they overreact—I can't promise that won't happen. But if we deal in facts and logic, calmly and clearly communicate the reality of the situation and focus on a solutions-orientated approach to the problem, those people will soon back off. As they say in the mafia, "This is the life we've chosen." The same applies to project management careers. When somebody recently out of school takes that first job in project management, they really don't know exactly what they're getting into. But for seasoned professionals—we know what this career comes with. Learn how to live comfortably in the chaos.

SOMETHING EXTRA—WORKING ON GLOBAL TEAMS

I want to touch on global teams quickly, because working with them, and on them, has been one of the highlights of my career. In addition to having traveled to places like Shanghai, Frankfurt, Argentina, Mexico, Thailand, and Toronto for business, I've worked with extended team members from many different countries. I find it really interesting to get to know different ways of life and working styles around the world. Each group has some things that I need patience to deal with, and things that I'm really impressed with and try to incorporate into my own style or process. It'd probably be seen as stereotyping for me to point out the pros and cons of working in certain areas of the world, so I'm going to try to avoid that—but, here are some tips for dealing with global teams:

Get to know each other: Some countries are better at this than others. Whether I'm working with process-driven hard-ballers or people who are less concerned with adhering to a strict process and more concerned about maintaining a stress-free environment, it always helps to be nice to one another. Really—it's just about being nice. Ask them about their families, or how their weekend was. We all like to have fun on the weekends!

In-person kick-offs: I was once lead PM on a global effort to change branding strategy throughout the world, and we had team members who lived all over the place. I had to bring together various support teams working out of Brazil, Mexico, America, and Argentina—but the client couldn't get the budget together for an in-person kick-off meeting. Plus, things were happening really fast and getting visas quickly was out of the question for the countries we wanted to travel to. We mustered through using web conferencing, but it was a mess for a while, until we were finally able to get some key people together in person. Sending one or two people to meet and work with others in person makes all the difference in the world, and I highly recommend it. Once you get to know someone, and possibly raise a glass together, it's much easier to deal with conflict if/when it arises.

Also, when you do travel to another country, get to know a few key phrases before you go. Don't fall for the hoax that English is the global language and everybody speaks it—maybe so in the office, but not in a taxi or in a restaurant. In addition to *please* and *thank you*, I usually try to learn these simple statements before leaving for a new country:

"How much?"
"Too expensive!" (I like to shop)
"Do you speak English?"
"Do you have an English menu?"
"Where is the bathroom?"
"Turn left [or right]"

Try to learn the numbers one through ten, and always have an in-language copy of your hotel's address that can be provided to a taxi driver.

Know the time zones: There have been times that I'm trying to coordinate so many time zones that I get really confused about this country being six hours ahead but that one is two hours behind … it's tricky to get a meeting together. There are a couple different things I do to help with this. First, Microsoft Outlook allows users to view more than one time zone on their calendar—so that helps quite a lot with setting up meetings. Also, sometimes I'll print out a list of what time it is in the other countries at every hour of our day. If you're working with people in Asia, get used to the fact that most meetings will either be at 8am or 8pm—after all, they're on the other side of the world. Hopefully one of you is a morning person and the other happens to be an evening person.

WebEx works: Working on global business for several years got me in the habit of attending all meetings via a web conference. Even when my global project was over and I started working local again, I forgot to attend meetings in person even though they were happening right down the hall. People were asking, "What are you doing? Come in here!" We can do amazing things using online conference tools—and the thing is, whether we're working with an agency down the street or in another country, it's the same experience. I really like to have the kick-off in person for large, lengthy projects, but if that's not possible then web meetings are just fine.

In this chapter we:

- ✔ Started the schedule
- ✔ Learned what activities take place during various parts of the schedule
- ✔ Determined the sprint groups
- ✔ Held an internal kick-off meeting
- ✔ Created a status report

5

INFORMATION ARCHITECTURE

I love the term *information architecture* because it explains the role just so clearly. Some people view an information architect (IA) as a resource who knows how to generate sitemaps and wireframes. Others may think an IA is a resource who understands customer's wants, needs, and online habits and can transform that knowledge into the perfect website experience. Well, those are both true, but what they also do is to ascertain the communication objectives of the client, along with their desired outcomes for the customer's online experience, and organize (or architect) the information to achieve those goals. That's a lot of stuff to know—they're smart!

At this point in the process we've been working with an IA to conduct some preliminary research, create the gap analysis, and put together the business requirements. Perhaps our research went as far as conducting some focus groups with select types of customers (the type of people who would consider a spaceship purchase) to get a better understanding of their preferences. At any rate, the IA now has a complete understanding of the project and is ready to make their recommendation on the new site structure. The first step is to create the sitemap.

SITEMAPS

The sitemap is one of those deliverables in the critical path that needs to be completed before the next step can begin. It may change as the project goes on, but it gives a good sense of the entire scope of the site. I hate to state the obvious (not really, I do it all the time), but the sitemap *maps* out each page of the website or application, indicating where each page will live in relation to the main navigation. We could even call it a graphic outline.

There are two types of sitemaps, although they're very similar. In this chapter we're not talking about a list of pages which can often be seen at the bottom of live websites to help the user easily jump from one page to another. We're talking about the graphic representation of what a site will become (see Figure 5.1). Sitemaps are created by the IA and are important for the following reasons:

1. They're a tool the creative and business teams use to refine their communication strategy. These teams can easily view all of the information on the site, where each page sits in relation to the drop-down navigation, and what other content surrounds the page. Looking at this visual site structure helps them decide if they like the way the information is organized or if anything needs to be added or moved around.

2. Sitemaps also help the team strategize about the customer experience. They can plan out the sequence of pages and information that customers will view as they click through the site, hopefully leading them on a path to a sale or other desired outcome.

3. The way the pages are visually represented on the sitemap provides graphic representation of which pages are applications or forms, if they link to internal or external pages, or if they're simple content pages. Notice in Figure 5.1 that application pages are represented by black boxes, and pages that direct users to an external site are shown as dark gray boxes. It doesn't matter how the pages are shown visually, we just need to make sure that there's a key somewhere on the page indicating which is which. Very often I've sat through sitemap reviews and have had clients ask questions like, "What content will be on that Careers page?" By looking at the sitemap, we can quickly determine that this page links to an external page which means we don't control the content of that page.

4. Another very important reason to make a sitemap is to present the global footer and header navigation strategy. The sitemap is now acting not only as an outline, but also as a tool that the IA uses to create and think through the strategy. In Figure 5.1 the

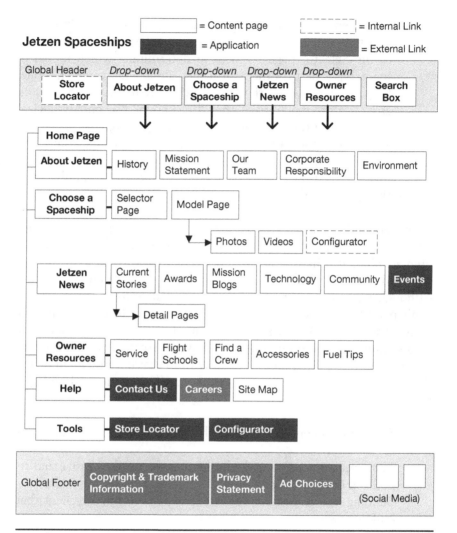

Figure 5.1 Sitemaps are not design tools but are used to organize the content

global header and footer are represented in the light gray boxes on the top and bottom of the page. We call them *global* because they'll stay static throughout the entire site—always available as an anchor for the user to see where they're at and how to get around.

5. Sitemaps also show the hierarchical relationship between pages and content. For instance, according to the sitemap we've created

for Jetzen, *Choose a Spaceship* is a category in the main navigation that will have a drop-down menu. The user can either go to the Selector page for help in determining which model of spaceship is best for them, or if they already know which model they're interested in, they can go directly to that model. Upon arrival to the model page, they'll be shown a secondary navigation allowing them to choose to see the photos, videos, or go to a spaceship configurator tool for that particular spaceship. The secondary navigation has a hierarchical relationship with the model page they're associated with, so the user can only get to these pages if they click on the model first.

Sitemaps are the first real deliverable within Plan and Define, and they set the stage for all forthcoming activities. Everything relates back to the sitemap. Since it's the first tangible deliverable, and in the critical path, it should be updated if any changes are made throughout the process. Although we're using a mobile first methodology for this project, we don't need to do anything special to the sitemap to account for the responsive design. Of course, if we wanted a different experience on the desktop site than on mobile, we'd have to create a version *B* of the sitemap, but we really want to stay away from that because it eliminates the benefits of being responsive. What are the benefits of building a responsive site?

- The same code is used to publish a site for many different screen sizes
- The same content is used, eliminating the need to resize and re-position images and text
- Building one responsive site is less expensive for the client than building two
- It's easier to maintain, since we only have to adjust the code or content in one place
- It's quicker to deploy since we do both at the same time

One thing to be cautious of when moving through the sitemap process is that some people on the team may try to gloss it over. Remember in grammar school when the teacher made us write outlines before starting on the paper? I remember thinking, "How can I know what the outline is before I write the paper?" I wasn't that great of a student. Of course now I can look back on this and see how backward my plan was—but this is actually how some people still approach the sitemap. They want to rush on to the next step even though the sitemap isn't quite sorted out, but we need to try to stop this from happening, if possible. The sitemap is a top-level view of the entire journey, and if one part of the map changes it's likely that other parts will need to change along with it. If we move on before the sitemap is done, then we'll end up doing double the work.

Once the sitemap is complete and approved by the client, we can start working on both the technical solution strategy and the wireframes. Let's start with the technical solution, because whatever we come up with there may affect the wireframes and functional specifications.

TECHNICAL SOLUTION STRATEGY

Also referred to as the *architecture diagram*, this endeavor attempts to map out all the data sources, integration points, and outputs so that a sound and secure digital solution can be created, maintained, and expanded upon. Again, the solution strategy should be done either prior to or at least during the wireframe process because the resulting diagram will provide the flowchart for data exchange and integration throughout the site.

This is the most technical segment of the book, and I won't go into too much detail. From a project management standpoint, our responsibility is to have a basic understanding of what a solution strategy is, who can provide it, and why it's necessary. Think of the solution strategy as a blueprint for the technical side of the project. In the home construction industry, a blueprint is a diagram the architects use to communicate to builders exactly how to construct the house. It maps out the electrical, plumbing, and HVAC systems among other things—and it is very similar to the technical solution strategy that we need for our website project.

There are two ways to approach the solution strategy. The team is either adding to an existing solution or building a new one. For the Jetzen project, there must be a solution already in place since they have an existing website. We just need to understand the current architecture, connect to it, and add to it. The team designing the solution strategy will need to review and understand our project's business requirements and functional requirements along with the client's current architecture structure that they'll be connecting to. A sitemap should also be presented.

Remember the old *Schoolhouse Rock* song called "I'm Just a Bill"? Maybe not, if you were born after 1975 or didn't grow up in the U.S., but I always loved that song because it described how a bill becomes law in the United States. It starts from the very beginning when the bill is just an idea in some citizen's head. Then that person calls their local congressman and starts the long, tedious process of *getting* a bill into law. The song goes through the entire process from idea to actually being a law, and we need to track our data in a similar way. In other words, track our little *bit* friends (see Figure 5.2) all the way through from beginning to end. A lot of this should have been started for us during the business requirements phase and now we just need somebody to build the plans for mapping it together.

In order to get a thorough understanding of the process, we need to define who the various data users and administrators are, then identify their job positions and responsibilities regarding the data—starting from where the data (bits) originates, all the way through to where it needs to end up. Of course throughout this same process, we need to know all of the current systems and applications in place for the data to be housed and transported. This is when we call upon the data architect who'll probably work with the infrastructure team and application developers to produce the data strategy. For an example, let's use something simple like a contact form (see Figure 5.3). The form comes in and—wait, where does it *come in* to? Does Jetzen have a corporate data center at headquarters, or should the form get routed to the local zone office based on the customer's zip code? Maybe the answer is *yes* to both questions. All forms come into HQ and then, based on zip code, get routed to the correct zone office.

Figure 5.2   Track those bits!

Figure 5.3 This simple form produces a lot of work on the back end

Well, wait, I guess it's actually not that simple. True, all the forms come in to the corporate office—let's call this database Receiver X. Then we send them along to the appropriate zone office, right? The form has a drop-down menu for inquiry type. Hmmm…well, what if their inquiry type is something that can only be handled by a specialty group like the parts department or the legal department? Well, now we have to figure out how many specialty groups the data might go to, and then figure out a way of automatically sorting the data into each of those groups. Then, like a pipeline of plumbing, we have to affix some sort of apparatus onto Receiver X to move the data into another pipeline so that it's directed to the right location (see Figure 5.4). Finally, we have to think about how each of the destination locations will connect to the data and work to communicate with each of those receivers.

Most of the connections I just described will be handled by Jetzen's internal IT department and we'll just need to connect with their system, but understand that it's not always as simple as it looks. As project managers (PMs) we have nothing to do with any of these decisions and are normally not even brought into the conversation, but we need to know that this is a step in the process—and one that we'll need to initiate. All of these decisions and challenges that I mentioned were based off of what appeared to be a very simple contact form. Think about what they'll have to go through for the news feed.

The output from the technical solution strategy is a diagram and, quite possibly, some written directions for what is happening on the diagram. Like I said before, this is a very comprehensive subject, but this book is not about designing data interfaces. Figure 5.5 shows a very basic diagram

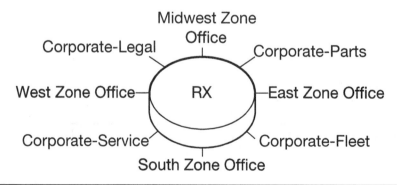

Figure 5.4   Incoming data should be as automated as possible to avoid long hours sifting through and directing communications

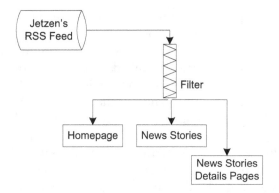

Figure 5.5 There's a lot going on in that filter—the data coming in from the feed has to be sliced and diced to work for each element of the site

for how the News Stories feed might come in from Jetzen's RSS, go through several different filters and then the appropriate data is sent to several different locations within the site. Stories and story teasers need to go to the homepage tiles, the actual news page, and then the details pages for each story.

We'll have to create a diagram like this for every application or tool on the site, but once we're done with this step, we'll be clear to work on the wireframes and functional specifications.

WIREFRAMES AND FUNCTIONAL SPECIFICATIONS

There's so much to learn about wireframes and functional specifications—they're really very interesting subjects. Even though they're not our primary function, it is really important that PMs understand what their responsibilities are in this area. First, let's discuss what the difference is between wireframes and functional specifications. In general, wireframes give a basic layout of the page from the customer's viewpoint, while functional specifications provide the detail behind how things work. It's very similar to the difference between business requirements and functional requirements.

Wireframes are line drawings that represent the framework of each page, and provide explanation for how the user interface will respond to the customer when they interact with the website. The first question that needs to be answered when discussing wireframes is, "Why do we need them?" People who are new to web design may think it best to go straight into styling and comps and let the creative team determine what the page should look like. That would be true for print projects, right? Not really. In advertising, print ads often start with a pencil or pen sketch—and wireframes are very similar. Digital projects are an interactive experience between the brand and the customer, and wireframing is an opportunity for the designer to collaborate with the IA to produce a customer experience that aligns with the business requirements.

Information architecture marries the art of creative design with the science of content organization and customer experience. It's not only what an IA does but who they are as people. Normally very bright strategists, who also have a strong creative side, IA's work well with creative designers. Both want the best visual representation of their work, but IA's are also driven to ensure optimal business results.

Global Navigation—Header

Back in Chapter 4 we took the sitemap and broke it out into digestible sprint groups that the team could manage:

Group 1: Global Navigation and Homepage

Group 2: About Jetzen and Owner Resources

Group 3: Choosing a Spaceship

Group 4: Jetzen News

Group 5: Help and Tools

Starting with group one, let's take a look at some wireframes for Jetzen Spaceships. Navigation is an important thing to document because it's often global and will be used throughout the site by the customer to find their way around. In this case the term *global* indicates that the navigation, both header and footer, remain static throughout the site versus having different drop-down menus or footer information on each page. This is really where responsive design comes into play, and when it gets fun. Up until this point, responsive was just something we were going to do *at some point*. Well, this is that point. We're still going to start with global navigation—so let's take a look at what that looks like between a small screen and a larger screen. Since we're working with a *mobile first* methodology, we start with the small screen as shown in Figure 5.6.

For mobile screens the global header is quite condensed—containing a menu icon that expands down to show the available menu selections. Everything on-screen and in the menu is shown in a vertical format, even if we were to flip the device sideways. Something interesting we can do to experiment with how website navigation changes as screen sizes get larger or smaller is to start from a desktop computer and bring up a popular website that you know to be responsive. With the site showing, minimize it so the edge can be pulled left or right. Notice how the content changes along with the screen size as we pull and push the edge of the screen. Starting from a small screen size, pull the edge to make it bigger and notice that as we reach a certain size width the *hamburger* menu icon from the small screen changes into a more traditional set of drop-down menus normally seen on larger screens. Also, images and content may even appear or disappear depending on screen size, but we'll talk more about that in the styleguide section.

The four frames in Figure 5.7 show the desktop (large screen) drop-down menus for each of Jetzen's menu categories. Even though the

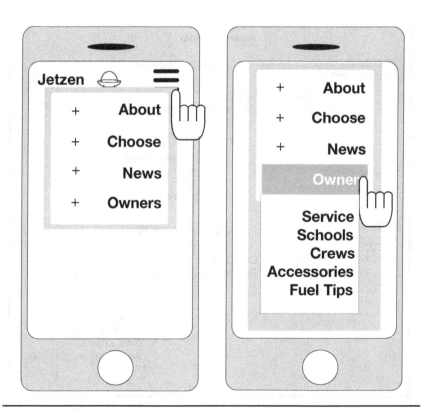

Figure 5.6 The menu icon in the upper right corner of the screen is referred to as the *hamburger*

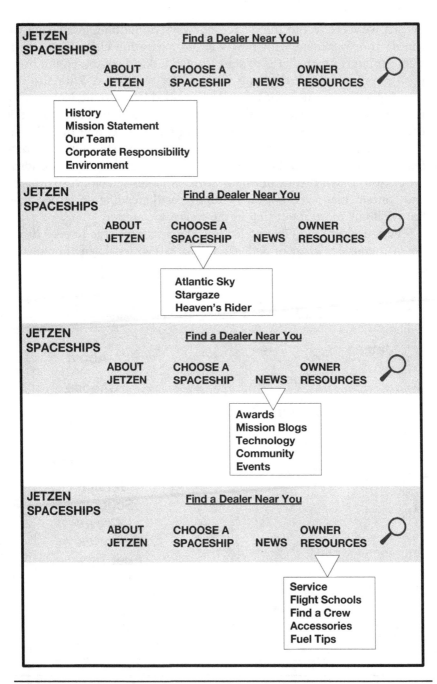

Figure 5.7   Check the navigation wireframes against the sitemap to ensure consistency

wireframe is not supposed to be used as a source to review content, I like to check these drop-downs against the sitemap just to make sure nothing has changed between it and our first group of wireframes.

The navigation will size itself according to the user's screen size based on the content strategy that our architects and designers will come up with, but there's a limit to this. For super-large screens there will, for sure, be a bit of white space between the edge of our website and the edge of the user's screen. At some point it becomes a poor user experience to keep spreading out the screen horizontally—information is easier to digest when the content is closer together. It can be quite tricky. There are a wide variety of screen sizes out there and we want our website to look great on everything, but there needs to be a careful analysis of the content and how users will view and interact with it. That analysis process will develop into a content strategy.

Jetzen's website will also utilize secondary navigation, which is a second row of menu selections located right below the main navigation. This allows the user to navigate around the particular section they're in without losing the main navigation, giving them the constant ability to easily relocate within the site. Figure 5.8 shows the mobile version for

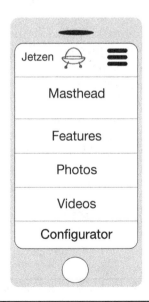

Figure 5.8 Upon touching, the spaceship icon is revealed to be *Find a Dealer Near You*

JETZEN SPACESHIPS		**Find a Dealer Near You**			
	ABOUT JETZEN	**CHOOSE A SPACESHIP**	**NEWS**	**OWNER RESOURCES**	🔍
FEATURES	PHOTOS	VIDEOS		CONFIGURATOR	

Figure 5.9   The desktop version of wireframes often uses underlines to indicate active page status

a model's features page, and Figure 5.9 shows the desktop version. For mobile, think of each of those areas as buttons that will either take the user to another page or expand to show content or even more available selections.

An underline is used in Figure 5.9 to indicate that *Find a Dealer Near You* is a hyperlink which would take the user to the store locator tool. *Choose a Spaceship* and *Features* use reversed text to indicate the active page (which page we're on). Remember that these are wireframes and that once the creative team styles the navigation those underlines may turn into something else, like a color change or some other graphic indicator.

Looking back at the sitemap for the Choose a Spaceship section (see Figure 5.10), I see something puzzling. Why doesn't the wireframe include the *Selector Page* we see on the sitemap? It doesn't seem to appear either on the main navigation drop-down or the secondary navigation under Choose a Spaceship. What the heck? Let's ask...

OK, I spoke to the IA and it was an accidental omission—he apologized and is working right now to update the wireframes. It'll be included with version two, once that's released. One might ask at this

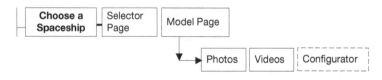

Figure 5.10   The Selector Page shown on the sitemap did not appear in the wireframes, so that must be questioned. Was is an accidental omission or did they decide to take it out and forgot to tell us?

point, what is a Selector Page? Well, we dropped that page in there because, to tell you the truth, most people don't already own a spaceship and aren't sure which one of the three available models best suits their needs. For these people we've offered a solution—the Selector Page (see Figure 5.11). This page provides a quick overview of each model and gives the customer the opportunity to determine at a glance which one they might be most interested in.

[Voice Over] "Just looking to tool around the neighborhood skies a bit on the weekend? Then the Atlantic Star is your best bet. Or, maybe you want to take a longer trip into orbit and show the kids the Ozone layer before it's completely gone. If that's the case, then head on over to the Stargaze page for more information. Speed demons will love Heaven's Rider—the most powerful engine in the industry."

As we can see from this example, the Selector Page helps direct the customer to the information they're looking for. But back to our wireframe discussion, there's another type of navigation I'd like to discuss, and its popular name is *breadcrumbs*. Probably upon reading those

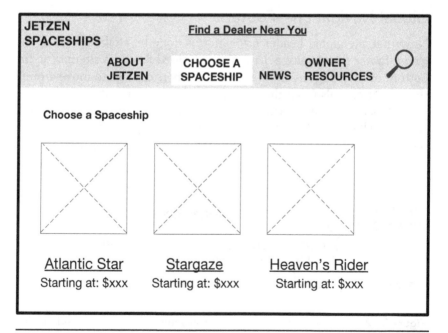

Figure 5.11 Model Selector Page

words the mind automatically went to a picture of somebody walking along in the woods leaving breadcrumbs behind as a way to find their way back. Wasn't that in a children's story or something? Anyway, that method doesn't work so well in nature because birds and other animals eat up the crumbs, but on the web it's a fairly effective way to show the user where they are. One common complaint from clients during a site review is that they have a hard time figuring out where they're at within the site, and to solve this they often request breadcrumbs. I actually had no idea what they were talking about the first couple times I heard it, I thought they were ordering a salad or something, but then I saw what the IA team came up with as a solution and it became clear.

Figure 5.12 shows the breadcrumb technique used for the Choose a Spaceship page. Every time the user clicks deeper into a set of pages the navigation dynamically changes to reflect their trail. In this example the user clicked on the Atlantic Sky model page under Choose a Spaceship, then clicked on the Configurator, and then on Exterior Color. Maybe they want to make sure the Atlantic Sky comes in their favorite color before heading over to the store.

Global Navigation—Footer

Now that the global header navigation is done let's take a look at the global footer (see Figure 5.13). We've decided to add a sitemap at the bottom of each page as another tool customers can use to move around the site. Notice also that the appropriate legal statements as well as the social links are located in the global footer—as expected based on the sitemap. All of this detail will appear on every page of the site. A small screen probably won't contain the sitemap part—that's just too much

Figure 5.12   Based on the breadcrumbs, can you tell if the user visited the Selector Page?

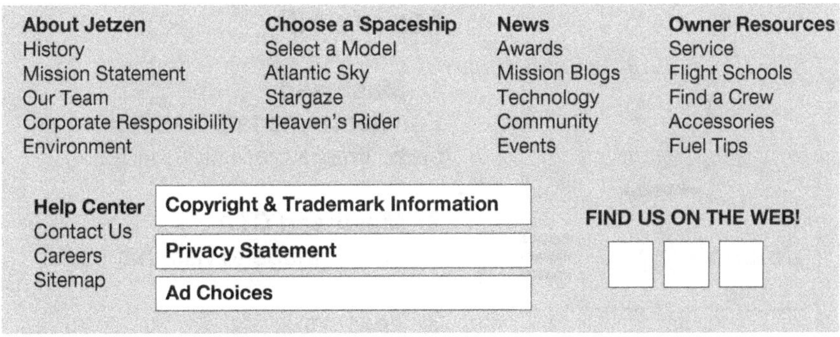

About Jetzen	Choose a Spaceship	News	Owner Resources
History	Select a Model	Awards	Service
Mission Statement	Atlantic Sky	Mission Blogs	Flight Schools
Our Team	Stargaze	Technology	Find a Crew
Corporate Responsibility	Heaven's Rider	Community	Accessories
Environment		Events	Fuel Tips

Help Center
Contact Us
Careers
Sitemap

Copyright & Trademark Information

Privacy Statement

Ad Choices

FIND US ON THE WEB!

Figure 5.13 Like the global header, the global footer may be seen on each page

information for that size area, but it will certainly contain the legal information and social icons.

Homepage

When we start to see wireframes for individual pages or applications, desktop computers generally have a wider variety of functional and visual *options* available, so the details provided along with the wires will also be more robust. Take our Jetzen mobile homepage for example (see Figure 5.14). We know how to operate a mobile device—we either click on something or slide our fingers up, down, left, or right—got it. Unless we are describing how *Find a Dealer Near You* uses global positioning or something like that, the general method of operation is understood and little detail is needed.

On the other hand, Figure 5.15 shows the desktop version of the homepage wireframe and has much more detailed functionality listed because there's more to explain when there are so many more options to consider. For instance, there are little arrows on the masthead and News Story teasers explaining that the images scroll, and also call-to-action buttons on the masthead that can be clicked. We also see that the global footer sitemap can be turned off or on by the website administrators. These are just a few brief details—some websites are highly functional with multiple items that can open or close, slide, pop open, or spin. It's important for the IA to annotate every detail regarding user actions and user interface responses. These annotations should clearly indicate how screen elements are intended to react to users and how the experience will use motion and graphic elements in the design.

Mobile HP

1. Masthead
Can accommodate for scrolling or static images, or link to video.

2. Masthead CTA
Click any image to visit that page.

3. Teasers
This area allows for news story teaser images that link to any interior or exterior page.

Figure 5.14   Wireframes should have descriptions for the page functionality

Besides going through the callouts, there are a few other things I'd like to call your attention to concerning the homepage wireframes:

1. None of the boxes on the page list their sizes. This information will come in the styleguide.
2. Paired with the drop-down menu wireframes we went through earlier, once the homepage wire is done, the team should have a very good understanding of how the entire site will eventually shape up and how it will be structured.
3. Although wireframes don't contain final text or images, it's a good idea to throw some content from the current site in there

Homepage

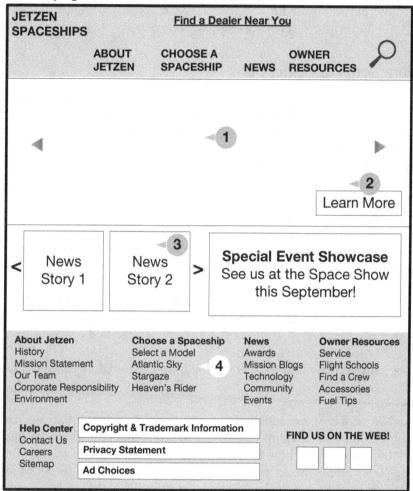

1. Masthead
Can accommodate for scrolling or static images, or video.

2. Masthead CTA
Call-to-action (CTA) buttons link to any interior or exterior page.

3. Catwalk
This area allows for news story teaser images that link to any interior or exterior page. Ability to scroll through several teaser images.

4. Global Footer
Provides agency functionality to turn Global Footer on and off, but it must be either ON all pages or OFF of all pages.

Figure 5.15   Read through the callouts—do they do a good job of describing how the page operates?

to help the clients really connect with what the team is building (like we did with the Special Event area). This helps the black and white wireframes come to life.

Functional Specifications

There's a fine line between functional requirements and functional specifications and the difference isn't all that easy to explain. One reason is because people use these terms differently depending on what type of project they're working on or what type of system they're working with. Here's the original description I gave for functional *requirements* back when we were reviewing the business requirements document (BRD): "Functional requirements give the detail behind the requirements. What happens when you hit the submit button? Where will the form be sent once the customer fills it out? How should the news stories be filtered and presented on the page?" Listed below are some requirements we could add to the homepage masthead:

- Hovering over the Learn More button placed over the masthead will change it to the version B image.
- The Learn More button can be placed at specified x/y coordinates based on creative direction.
- Images will automatically swipe from left to right every eight seconds.

And so on. Those are functional *requirements*—objectives that we have to meet in order to fulfill our agreed-to scope. Functional *specifications* are the rules behind how to make those requirements operational. Take the News Stories teaser tiles as an example. What do we know so far? Let's review just one of the functional requirements that were recorded in the BRD along with some specification details we need to solve for.

Requirement:

- News Stories should be sorted into at least one of six categories (Awards, Community, Mission, Blogs, Technology, Events)

Specifications:

- Sorting is a manual process that will require each story to be tagged into one or two of the categories

- The feeding system must be able to pass through the following information
    - ▼ Teaser copy
    - ▼ Full story
    - ▼ 560 × 250 maximum width images
    - ▼ Category tags
- The receiving system must be able to capture all of the previously listed information and also filter by category
- Once filtered, the system needs to use established rules that will send each story to specific locations within the site according to category
- How will the stories be pulled into the homepage?
- How will the images be displayed compared to the teaser text?
- How do we ensure that a good mix of categories is represented on the homepage?

Of course that's just a short list, and it is beyond the scope of this book to attempt to go through and come up with actual specifications for this fake project. I just want to communicate what the steps are. It took me forever to figure out functional specifications, and to not be surprised by needing to get them done—and I mean *forever*. So many times we'd wrap up wireframes for a project I was working on and I'd tell the client that we were ready to start development, but we weren't. We still had to do the functional specifications! What? There's more? Ugh.

Now that the functional specifications are complete we need to bump them up against the technical solution to make sure it all still works together. The wireframes and functional specifications are both inputs to the styleguide—and styling can't start, or at least can't be finished, until both of these inputs are approved. Please make sure the wireframes are routed and formally approved by the client, so we can send this sprint package off to the art director in charge of creating the styleguide. Remember our sprints? Now the IA can move on to Sprint 2—while Sprint 1 is being styled.

STYLEGUIDES

Wow, we're already talking about the styleguide. We've covered a lot of ground! How about we pause for a moment and check our process to see where we're at? (See Table 5.1.)

Table 5.1   Process map with check marks

PLAN AND DEFINE	
INITIATING	**PLANNING**
✓ Gap Analysis	✓ Plan and Define Schedule
✓ Workshop	✓ Communication Planning
✓ Stakeholder List	✓ Sitemap
✓ Business Requirements Document	✓ Technical Solution Strategy
✓ Preliminary Budget Estimate	✓ Wireframes and Functional Specifications
✓ Statement of Work **(Tollgate)**	12. Styleguide
	13. Analytics Analysis
	14. SEO Analysis
	15. Infrastructure Assessment
	16. IT BOM
	17. Development and Change Management
	18. Test Strategy and Cases
	19. Package Identification
	20. Construct to Close Schedule

CONSTRUCT TO CLOSE
21. Create Content Tracker
22. Asset Quality Review
23. Content Entry
24. Quality Assurance
25. System Integration Testing (SIT)
26. User Acceptance Testing (UAT)
27. Nonfunctional Testing (Performance, Security, Disaster Recovery, Failover)
28. 301 Redirects
29. Cutover Management
30. Transition to Operations

Oh my goodness, that doesn't look like a lot of check marks from this angle, does it? Thirty steps? Every time I start a new chapter I tell myself, "This will be the hardest one—after this I should start to fly right through it." But that never happens—I never fly. We're not even halfway through yet. Honestly, let's tackle the styleguide, and then I *really* think it will start to ease up (yeah, right!).

The Grid System

According to our resource discussion, both the creative design and style-guide are handled by art directors, but the styleguide designers have specialized skills for transferring fonts, colors, and layouts into a language that coders can work with. My friend and colleague Melissa Knisely is an art director who specializes in creating styleguides, and she told me, "Styleguides specify what the standard headlines, sub-headlines, paragraph text, forms, buttons, links, etc., all look like and what their transition behaviors are. It also specifies page background information, layer orders, and asset usage." The first thing she nails down is the grid system, because it provides the framework for both the creative team and our developers to follow.

The grid system really only applies to desktop, so we're not talking mobile right now. Mobile first activities happen mostly in the design of the website, which we've already done. At this point we're documenting the specifics of the design so that the developers know what they're supposed to do. Plus, we don't need a grid for small screens—they have one column and the width measurements are relative percentages of screen size.

I'm sure art students everywhere are horrified the moment they learn that math is a requirement in art school. Wait, math and art don't really go together, do they? Yes, they do, but in web design pixels are used as units of measurement instead of inches or centimeters. To conceptualize this, think of the computer screen as a piece of graph paper. The lines cross each other at x and y coordinates on-screen just like they did on paper. We're going to use the x and y coordinates to come up with our columns and grids for the desktop site, but we really don't lock our images and content down to those positions or the site would not be responsive.

During the grid exercise the creative team and the IA teams decide together on what the grid system should be by putting different column examples in place and figuring out what the pixel width should be per column, for the gutters and for margins. Figure 5.16 shows a typical twelve-column grid system. Notice the twelve gray vertical columns in the background and then how we broke them up into sections with the white horizontal rows.

So how does this apply to our website? Well, if we cut an image to put in the masthead we'd make the maximum width 960px wide. If we needed to put two images side-by-side, they'd each be 470px maximum

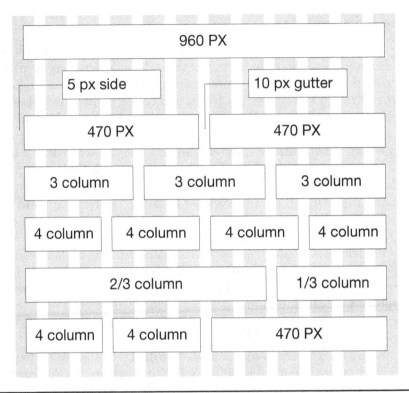

Figure 5.16 The grid system provides structure and consistency

width with a 10px gutter in-between and 5px between the edge of each image and the side of our live area. Based on our homepage wireframe (refer to Figure 5.15), are you able to figure out how many columns we'll need for the news stories and special events row? At first glance one might say *three*, but it's actually four. We start with a four-column grid and then merge the right side together to make three. This couldn't be done starting with a three-column grid.

Our website is responsive, so those images are going to grow and shrink according to screen size, or what is also referred to as the *viewport*. Exactly what the designers want to happen at each interval is based on the content strategy previously put together. The content strategy is then implemented by using some of the techniques in the following list which are written into the Cascading Style Sheet (CSS) code. Note that we can use all or just some of these techniques—they all work together to manage the content strategy that our IA's and designers have developed. There are more out there as well, but these are the ones I hear referred to most often.

1. **Relative units:** This means that at 960px, the image we're using in the masthead is at 100% of its maximum width. When coming down from a large screen onto a small one, we can still see the entire image and nothing is cut off—although, to make this happen we'll have to come up with an equation so that, based on the user's viewport, images will be sized at a percentage of their maximum width.

2. **Breakpoint:** Working from big screens down to small screens, the breakpoint is the point at which images or text in the same row stop contracting and instead drop down to a new row and we lose a column. So instead of two images placed in two columns next to each other, they are now on top of each other in the same column. Where the exact breakpoints *are* located is determined in the content strategy.

3. **Max width:** Picture the same sentence shown on both a large and small screen. On a small screen it may look like a paragraph, while on a larger screen it's just a tiny line. Using *maximum width*, we can make sure the text or image won't expand past a certain width on a large screen, making it look more esthetically pleasing. Maximum width is also used with images for several different reasons, including:
    a. To work within the grid system
    b. To establish a base by which relative units can be calculated
    c. To limit expansion on larger screens which would produce a grainy looking image

4. **Vector images:** If you were following along with me on the 960px masthead example, you may have wondered how the same size image works for all screen sizes. Vector images are a file type which keep the same quality whether we show it on a mobile phone or on a large screen. They can even be blown up large enough to be used on printed billboards.

5. **Responsive content:** The developers can create rules in their code that can cause content to appear or disappear depending on screen size. This is commonly used when the desktop site is much more robust than is desired for a mobile screen, so we tag certain assets to either go away or appear based on breakpoints.

Cascading Style Sheet

After this initial grid layout is done the real work begins. The creative team sends over their design files and our styleguide expert starts measuring out each and every last pixel to determine the html language for each color, font, link, and image. The developers then use this information to create the CSS which drives how it all publishes on-screen. Now, please don't look at any of these examples too closely—I am just making stuff up here. For instance, in Figure 5.17 I'm trying to show how every pixel of the site needs to be measured out and fonts need to be specified so that the developers can grab that information and create their code.

Every button, every arrow—everything has to be specified. The tables underneath the artwork show how text will be handled based on whether it needs to be normal, or if it needs to indicate that the page has already been visited, or what the hover state is, etc. Figure 5.18 attempts to show a button in both *normal* and *rollover* states. This information also goes back to the original creative designers to ensure that their entire team is following the rules. There will certainly be some back-and-forth between both teams until all the kinks are worked out and everybody is happy with the final result.

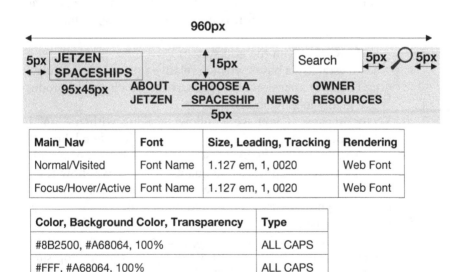

Main_Nav	Font	Size, Leading, Tracking	Rendering
Normal/Visited	Font Name	1.127 em, 1, 0020	Web Font
Focus/Hover/Active	Font Name	1.127 em, 1, 0020	Web Font

Color, Background Color, Transparency	Type
#8B2500, #A68064, 100%	ALL CAPS
#FFF, #A68064, 100%	ALL CAPS

Figure 5.17   Example of what you will find in a styleguide

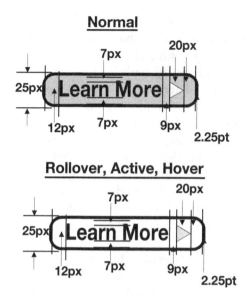

Figure 5.18 Styleguides are very tedious work

Main_Nav	Font	Size, Leading, Tracking
Normal/Visited	Font Name	0.0966 em, NA, 0025 em
Focus/Hover/Active	Font Name	0.0966 em, NA, 0025 em

Rendering	Color, Background Color, Gradient Fill, Rule	Type
Web Font	#FFF, #0EBFE9, #05B8CC Left, #FFCC11	Initial Caps
Web Font	#FFF, #8B2500, #A68064 Left, #FFCC11	Initial Caps

Suddenly, being an art director doesn't look like so much fun, does it? The styleguide for a large site can be 200 pages long and take months to create. That's why, if you look back to the budget, I have $80,000 in there for the CSS and styleguide, and that's pretty light! That has to cover both the art director who makes the styleguide and the developer who turns it into code.

In addition to the CSS, styleguides set standards for creative consistency throughout a site, so that every headline, subhead, or paragraph text looks the same on every page, as well as primary and secondary buttons, the color palette, etc. It's normal for several different designers to not only build the original site, but also to pop in and out for maintenance of the site for years to come. Having a styleguide in place allows everybody to review the established rules so that their work matches and fits in with everything else.

Of course having a guide brings with it some consternation at times. I think most designers would agree that styleguides are vital to the web design process—but they also provide limitations, obviously. For example, I can imagine that in the future Jetzen will probably *launch* a new spaceship model at some point, and the brand manager for this exciting new product might want their own exciting new look and feel. I've seen styleguide rules broken before, and although I may have personal preferences, it's not my job as a PM to decide whether or not the rules may be broken for any given situation. But, just as a warning, when a company implements a styleguide there has to be governance in place to make sure it's followed.

OK, that wraps up our styleguide for Sprint 1. Scope freeze! We made it! At this point, developers are going to start coding, so I always call *scope freeze* with the team and make sure they understand that any changes beyond this point may not be able to be accommodated, at least not during this phase of the project. If it's something pretty small (and something the agency may have overlooked), then we always try to bring it in. But if it's something brand new, we may have to give the clients the option to either move this new requirement to Phase 2, or add it to our current project with the understanding that the schedule and budget will change.

Although we always offer to deliver the styleguide to the clients, they'll have very little to no input. It's very technical and specific to the needs of designers and programmers. What I usually do to keep everybody in the loop is to have the person creating the styleguide present their accomplishments about once a week during the weekly status meeting. This way they can ask all of their questions with the right people in the room. And, there *will* be many questions going back and forth between the styleguide art director and the design team. Instead of trying to get in the middle of those discussions, I just ask them to exchange contact information so they can work together throughout the week. Then, any big issues that need a client decision can be brought up during the status meeting.

So what happens next? The CSS developers take the styleguide and start writing their code. Once they are done, things start showing up on the screen.

In this chapter we:

✔ Designed the sitemap
✔ Developed the technical solution strategy
✔ Created the wireframes and functional specifications
✔ Made the styleguide

6

ANALYSIS

In this chapter we're going to go over four process steps that don't need to be done in any sort of order other than after the styleguide, since at that point, requirements should be completely confirmed and we've entered scope freeze. In fact, they can all be done at the same time. I normally wait until the entire project (all sprints) are through styling before starting analysis because I feel it streamlines the process, and it helps the folks we'll be talking with to get an overview of the entire project, instead of just chunks of it. That way they can confirm their recommendations without any caveats pending their full review. Over the next several pages we'll be tightening up the reins and making sure we're not missing anything. Get ready to discuss:

1. Analytics
2. Search engine optimization
3. Infrastructure management
4. IT bill of materials

ANALYTICS

A lot of people get analytics and search engine optimization (SEO) confused. For a long time I didn't even understand what the difference was, so just in case you're having the same problem, let's start out with an explanation. SEO is what gets people to the site, while analytics is the study of what people do once they've reached the site. Here are some questions that analytics advisers try to answer for their clients:

- How are people arriving at our site?
- Which pages are they most likely to visit?
- How long are they staying on those pages?
- At what point are they leaving the site?
- Are they using the tools we developed, such as, the Configurator?
- Do they click the *Learn More* buttons?
- What is the typical experience, and does it change based on how they arrived at our site?

The list could go on and on, and it often does. Once we have code in place to answer the first set of questions, the team will come up with another set of questions they'll want answers to. Just a little bit of caution here, though, as analytics code is added to the site, it can also slow it down. I've seen companies with long-standing sites that have had so much analytics code built up over the years that they've had to go through and remove some, due to site speed issues. Once they're approved, the analytics team will want to see the sitemap and the wireframes so they can review these documents and create their tracking recommendations. Like any deliverable, they'll provide their recommendation to the client, and the client will review the material to make sure all of their key performance indicators (KPIs) are included. KPIs are strategic targets, established as a measurement of success.

Key Performance Indicators

KPIs often come down from top-level management and end up as employee goals which may result in *something awesome* like bonus money for the client if KPIs are trending upward or *something bad* if the KPI targets are not being met. If the company is not meeting its KPI goals, then there's something going wrong with the people, the process, or the product. Here are a few KPIs that might apply to the Jetzen site:

1. **Unique visitors versus returning visitors:** If the visitors are mostly returning guests, rather than new, then there may be a problem with the reach of the advertising. A company has to keep pulling in new potential buyers in order to stay successful.
2. **Percent of visitors who complete the spaceship configurator steps all the way through to the summary page:** Research shows that people who go through the configurator process are more

likely to purchase, therefore Jetzen wants to see a lot of people completing that process. It's hard to make the actual connection between completed configurations and sales, but they should both go up at the same time.

3. **Percent of visitors who attempt to locate a store:** Again, why would somebody want to locate a store if they didn't plan on going there to shop for and possibly purchase a spaceship?

See how these are very much geared toward sales and why they might be called KPIs? If these numbers are doing well, then that's a good indication that the company is doing well and sales should be strong.

Once the analytics recommendation is approved, we hand that over to the analytics developers to implement on the site. This is coding work, and specialized developers may be needed to do it correctly. Every once in a while the analytics team requests tracking that doesn't end up getting implemented. This sometimes happens either because making it work properly is impossible due to some other dependencies, or it's not worth implementing based on how long it will take the developer to implement, or it will create an unnecessary decrease in site speed. Usually these things come up after the developers have had time to review the tracking request. If the programmers didn't do this every once in a while, then we should be concerned—because in that case, they're just order takers and not acting as expert consultants. They know this stuff better than we do, so we need their expert opinions. When they come back with questions, concerns, or suggestions, just mark them down and provide the client with their options and let them decide how to proceed. The options may be to pay a little more money to get the analytics done properly, or spend a little more time, or to just drop it altogether. These are decisions the client has to make—after all, it's their site—we just provide them with all of the information they need to make an intelligent decision for their company.

During the testing phase, the quality team will make sure that the analytics are firing correctly before allowing outside parties to perform their tests. The test team can check tracking through open source add-ons which provide debugging, editing, and monitoring of website code. These tools show up on a user's desktop and show what's happening behind the scenes as we click around the site. Once the internal test team approves that everything is working correctly, we can give the clients and analytics team the go-ahead to check their report suite dashboards to make sure the reporting tool is receiving the data as expected and showing up in their reports.

✔ Check mark for analytics

SEARCH ENGINE OPTIMIZATION

SEO is the creepiest subject we'll talk about. There are spiders and crawlers and robots and entities and all sorts of creepy stuff. Plus, it's totally politically incorrect with all of the classifications, rankings, prominence factors, and trust issues. Where to begin? Probably at the beginning, I guess.

Most people today understand what SEO is, but just for safety's sake, I'll just lay it out there. Websites can be optimally built so that search engines are more apt to find them and rank them when people search for terms that match the content on the site. For instance, if I were to search for *spaceships*, I'd really hope that the Jetzen website would come up not only on the first page but as number one on the list. When our website appears first through third on the indexed list, this promotes more traffic to Jetzen's site, more customers shopping Jetzen's spaceships, and more sales.

Of course clients can always use paid search—where they purchase certain terms (like spaceship), guaranteeing an up-front ad listing. But what we're talking about here is natural search, which are search results that are ranked based on the number of clicks to that site, relevant content, and click-throughs to other pages. Natural search can also be called *organic* or *non-paid*, and most corporations use a combination of both paid and natural search to guarantee ranking.

Now, the problem with this chapter is that I'll write it based on what's causing page relevance today, but by the time this book is printed, these things could change. They probably will change, actually. Remember the main thing I want to get across in any of these chapters is not the *how* but the *why*. Why do we need to make sure SEO is a strong consideration when building the site? Because if we don't make the site easy to find, there's a good chance nobody ever will. And it's not easy work. A friend of mine left the SEO business and went into something else altogether because he said every time the team would get the SEO up-to-date for a site, Google would change their algorithm and they'd have to do it all over again. Bummer. Why does Google do that? I don't know—Google it.

Meta Tags

If there's a computer or tablet available as you read this page, go to a website that you think is pretty credible—a corporate or manufacturer site that you like. Now *right click* on the page and select *View Page Source*. At the top of the source page are a lot of lines that begin with the term *meta*—meta description, meta keyword, meta title, etc. These lines show up on every page of the site (or they should). Here's an interesting fact—look at the meta description from the page source and then go to a search engine and make the website you are using for this exercise come up in the search results. Notice that the description the search engine provides is the same as the meta description.

Content Optimization

In addition to implementing the meta data, we need to make sure the content on the site is optimized as well. When search *bots* crawl websites for content, they're looking for codes such as H1, H2, H3, and P. Here's what those codes mean:

H1—Main headline of the page which contains the most weight regarding page content

H2—Sub-headline of the page, falling just below H1 in priority

H3—Indicates a third tier headline

P—All paragraph text is tagged as a *p*

When search bots read these codes, they know that the H1 tag is the most important copy to look at on the page, and should provide a good indication of the page's content. After finding the H1, the bots look for the H2—weighing those words as a bit less than the H1 copy, and so on. Sites are crawled and re-crawled periodically to get an update on site content. The keywords and phrases the crawlers find will then get connected to search terms, and indexed in order of relevance. The search engines are our friends—they just want to know what content our site has to offer so they can, in turn, recommend the site to consumers. Now that we have a general understanding of how SEO works, let's get down to what we need to do for the Jetzen website.

The first thing that usually comes up regarding SEO is the schedule in regard to asset hand-over. An asset can be any image, text, video, or other piece of content that we're going to put on the site. If the site provides downloadable PDF files, then those are assets. All of these items fall into the category of *things the creative agency provides to the production agency*. Here's the process I like to follow with asset handover (see Figure 6.1).

The SEO team reviews all assets prior to delivery to make sure they are optimized for search. Here's what we can expect from their team.

Image naming conventions: These are the actual names we give to the images. For instance, when the creative agency delivers the Jetzen logo (I believe the styleguide said it should be sized 95 × 45px) we need to know what they want us to name it, since the search engines can read image names. They'll probably tell me to name it *2015jetzen-spaceships .png* or something like that. The key here is to get this list of naming conventions at the same time that the actual assets arrive so we don't have to name them twice. I've been on projects when the image naming conventions were not delivered on time by the search team, but we wanted to get started building the site, so we named the images ourselves just to get things started. Then, once the SEO arrived we had to rename them all. Not the best use of everyone's time.

Image, video, and flash alt tags: Anywhere the creative team is going to place an image, video, or flash file we need to incorporate alt tags. To learn what an alt tag is, go to any website on a desktop computer and hover over an image with the mouse. Chances are that after a second or two some copy comes up that seems to describe the image. This is alt text. Obviously search engines can't *read* a video or picture, so we use alt tags to provide them with descriptions. Here's an interesting

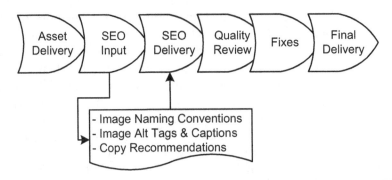

Figure 6.1   The assets are first delivered to the SEO team for their input, and then the SEO team delivers those same assets to the production team for asset quality review

fact about SEO related to images, videos, flash components, or anything else that can't be read by SEO. The concept of having words in place (alt tags) to describe what's showing on the screen was first created to provide an optimal user experience for the blind. Blind people utilize *automated screen readers* when they're surfing the internet. The screen readers pick up the alt tags and read them to the blind so they have an understanding of what's showing on-screen. So, not only do alt tags provide SEO enhancements, but they're also a helpful resource for the overall community.

Image captions: Image captions are pretty easy to understand—I have used them throughout this book, but obviously not for SEO reasons. Because copy that's baked into an image isn't searchable, we need to use alt tags and image captions to provide the search engines with something that identifies the content so they can match it up to what people are searching for. It's also super helpful to the viewer.

URL: Every single page on the site will have a unique URL, and we are able to determine how the URLs will read. The only thing we really can't change is the domain name. In our case the domain name is going to be *www.jetzen.com*. Let's take a look at some pages associated with the Stargaze model section (see Figure 6.2). Remember what that looks like?

Based on the sitemap, I would imagine that the page URLs for this section under Stargaze might look something like this:

www.jetzen.com/model-selector

www.jetzen.com/stargaze

www.jetzen.com/stargaze/photos

www.jetzen.com/stargaze/videos

www.jetzen.com/stargaze/configurator

Copy recommendations: The search team's job is to understand how people search for things they're looking for. In this case they'd be looking for spaceship sales, services, or accessories. They have lists and lists of keywords people type into search engines when looking for spaceship-type things and they like to have the copy writers drop these words into their text. Now, this gets interesting. Here we have two somewhat opposing groups trying to work together. The copy writers are creative individuals who aspire to write words that are engaging to the customer, while the SEO folks are more results oriented and want to use keywords that may not sound great but get better search results. I can see both

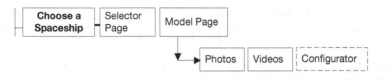

Figure 6.2   Each spaceship model will have a unique set of URLs

sides—and I'm not now, nor will I ever get, in the middle of it. I'm just saying it gets interesting.

The best way to deliver the SEO recommendations is through the copy document (*copy doc*). When the writers turn over their copy docs to the SEO team, the SEO team needs to find each image (videos, flash, PDFs, etc.) called out within the doc and provide alt text, captions, and naming conventions so that the production team can use one document to get all the information they need. Also at a page level, they need to provide the desired URL and meta information. I've tried it other ways, too. For instance, at one point we were creating an excel file that went along with each copy doc. The excel spreadsheets contained all of the various images and whatnot in one column, and then had columns for the alt text and naming conventions, and so on. That was much more difficult than just including everything in the copy doc. The copy doc method was much easier for everybody involved.

So where am I going with all of this? It's important to engage with each of these parties and explain how the process is going to work so they can plan for it. We don't have to explain the importance of SEO to anybody on the team—they're professionals and will already know. What we may need to explain, however, is the process for handing over assets and how the SEO team will submit their recommendations. It's the last step before assets can be delivered to the production agency, so we want them to be ready and prepared to move quickly so they don't hold us up.

SOMETHING EXTRA—A/B TESTING

A/B testing isn't new but it's one of those terms that most project managers (PMs) wouldn't know the definition of, unless they had worked on it before. Plus, it's really interesting—so I thought I'd give a quick report on the subject. This testing takes place when the art directors and information architects have a couple of different ideas about how to design a certain user experience, but they don't have any research to support which version will appeal more to users. Or, more specifically, which version will be better at reaching the desired outcome. In this circumstance the teams will build both versions (the A version and the B version), put them both up for a certain amount of time, and measure which one gets better results, by working with the analytics team. When KPIs are being measured, the design is less often about how it looks than actual results.

An example of this for Jetzen might be if the team was challenged with getting more users to complete a spaceship configurator. The company might want to drive interest in their spaceships through this online tool which has been proven to be closely linked to sales. The team decides to put a new teaser on each model's features page directing customers to the configurator tool, and they're not sure which design to use. Which one will result in more click-throughs to the configurator? The difference between the two teasers might be that they use different colors, different verbiage, or they might alter the placement of a button. Since the team is unsure which one is better, and they know that management will be looking very closely at the results of this project, they might implement an A/B test to be certain.

MANAGING INFRASTRUCTURES AND ENVIRONMENTS

The third thing on our list to talk about in this chapter is infrastructure. There are several things in this category that we need to review before we're ready for development to begin on the Jetzen project. Our objectives for this section are as follows:

1. Ensure the correct types of environments are in place
2. Ensure the infrastructure can support this new project
3. Ensure the team has a release management process in place
4. Start the domain mapping and v-host configuration process

Infrastructure managers don't have the most glorious jobs in the world, but for sure they have one of the most important. The most crucial thing they do is to manage the code release packages from one environment to another. Our fictitious team works at a digital production agency, so I have to imagine there's a lot of code being slung around. With many different clients and projects all sharing the same infrastructure and available environments, there's bound to be some code that doesn't get along with each other. Chaos ensues, things break, clients panic, and we're up all night trying to fix it. This happens—but, one of the best ways we avoid missing our bedtime and eating random leftovers from the company fridge at 3 o'clock in the morning is to have a good infrastructure team in place to manage code deployments.

Managing Environments

Figure 6.3 shows a typical infrastructure of development environments that a production agency might use. Larger companies might have more environments and more structural levels to work with, but the system should be basically the same as what I am about to describe. The main reason for all of this hierarchy and intense scrutiny is to ensure the stability of the production environment—the live site. If things go wrong up there, again, nobody sleeps until it's fixed. Of course I'm not explaining all of this because I think it's the PM's job to set this all up, but it is our job to understand it. And, to tell you the truth, I've worked in a smaller agency where I had to explain this system to them and ask to have it put into place. Things are really going wayward when the PM is setting up the infrastructure.

For Jetzen we're working at a huge production agency and our infrastructure and environment teams follow a rigorous process to maintain stability and integrity. At the bottom of the hierarchy shown in Figure 6.3, notice development areas A and B. There could be any number of environments for the developers to use, and it's a good idea to have several available, because these guys and gals are updating their environments with new code potentially several times a day. If we didn't have several environments for them to work in, they would disturb each other with the constant updates (environments *go down* for a while when code is being updated) or possibly create code that accidentally messes up somebody else's project. It's a good idea to keep the projects as separate as possible.

The development area is where programmers write and test their code before anybody else can see it. It's their working environment. The code must be published in some way, in order for them to make sure it's working correctly—but since there's a lot of sandbox type activity going on in there, it's not a good idea to show it to the extended team. It's hard to explain to the account team to not worry about the flying seagull that sweeps across the page every 10 minutes—that's just Jacob

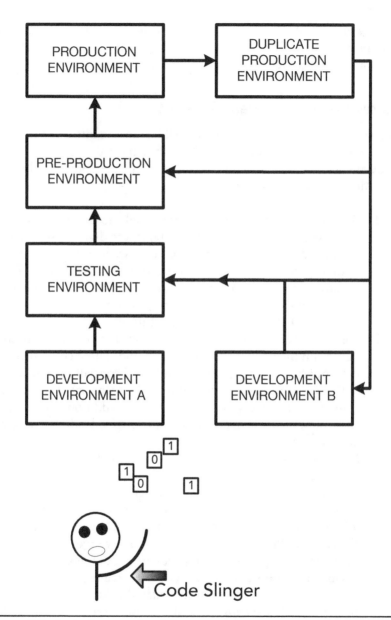

Figure 6.3 Lack of appropriate infrastructure management can lead to failure in the production environment

messing around. Well, you know what I mean. Once the infrastructure team provides us with an environment assignment we can get the developers started with their work. But before we do...

Hardware and Software Support

Once we know which environment we'll be moving into, the infrastructure team will want to make sure that they have the right amount of hardware and software available to support this new project. There are a few different things to take into account.

1. **Additional assets:** We'll be adding weight to the system from all the new assets that will be loaded into the digital asset management, and the infrastructure team may want us to provide them with an estimate of the quantity and size of these assets. This is usually done by comparing the site we're building to an existing site so that the infrastructure manager can get a ballpark of what they'll need to support in the near future.
2. **New users:** Since this is a new client, we may need to hire some new team members to support the brand. These could be either developers or content team members. Especially when dealing with a content management system, the team will want to make sure that the hardware can handle the additional stress from these new users logging in, building pages, loading assets, and activating new content.
3. **Site traffic:** Most importantly, the infrastructure team will need to make sure that they're ready to support the amount of site traffic, spikes and all, that are expected on the site. Even the traffic that isn't expected! Since Jetzen has an existing site, we'll ask them for traffic information for any three-month period over the last year. We'll want to know how many unique visitors, how many page views, and which pages are visited most often.
4. **Technical solution:** With the new site, we're adding on a few new tools that will require an updated solution strategy. All of this extra data exchange will also cause some stress on the system and we need to be ready for it. Go through the sitemap with the infrastructure manager and they'll determine what effect the new functionality will have on the systems.

Release Management

To explain the release management process, let's pretend that for Jetzen our developers are working on the *Find a Dealer Near You* tool, and have it ready for internal testing. Since our company has a highly regulated release process in place, we have to wait until next Thursday for our code to be available in the testing environment because there's another team in there right now finishing up testing on another project. If we were to sling our code up there in the middle of their testing period, then our new code might affect their test results. They could be testing out a new taco locator when we drop our code in there, and all of a sudden something goes wrong and tacos are nowhere to be found—anywhere. People are starving. It'd be difficult to tell if our new code messed something up or if their code wasn't working—that would be bad.

After waiting very patiently until Thursday, our code has finally been released into the testing environment. All applicable parties have been notified and we're in there testing the heck out of it. We only have a few days to test, though, because we have to enter all of our bugs into the ticketing system and give the programmers ample time to make the changes before next Thursday's release. We'll do this back and forth for a few weeks, until the tool completely passes testing and we're ready to move it up to the preproduction environment.

Once we're in preproduction, the extended team will start their reviews, so it had better be close to perfect. Also, this is the last step before the code gets moved up into production, and if there's any sort of problem at all, the infrastructure manager won't let us move forward. They won't risk the stability of other live sites for our little tool. And let me tell you these guys are under a lot of pressure. Every team thinks their project is the most important thing happening in the entire world, so the infrastructure manager has to be tough—and able to say *no* to people. Their job gets very stressful!

When everything is perfect and thoroughly tested, we can move up into production, according to the normal release schedule that's in place. Production releases are scheduled a lot less often than the lower environment releases—sometimes only once a month or so. Notice from Figure 6.3 that there are two production environments. One reason to have this duplicate area is for what we call *hot fixes*. A hot fix is an issue that's so crucial to the client's business, for legal or other reasons, that it must be fixed prior to the time the next production release is scheduled to deploy. If we have to put a hot fix into production, then either something important is not working right or something about the product specifications or pricing has changed and it's imperative that it be updated immediately. As you can imagine, when this type of thing happens people are in a rush to fix it quickly. They're under so much pressure to move quickly that they might accidentally skip a couple of quality gates in order to get the fix deployed, and the infrastructure team needs to do everything in their power to eliminate risk to the live site(s). When people are under extreme pressure from their clients, they can sometimes forget to make good decisions for the overall team, so the infrastructure managers have to remain calm and focused.

Infrastructure managers use the duplicate production area to test hot fixes. They don't want to put the fix code into preproduction because that area can be slightly unstable due to the release testing that's going on. Plus, as I've said before, throwing our code in there while other projects are being tested could be a detriment to the other teams who are testing in that environment. That's why we have the duplicate production environment. This area is exactly the same as production, so it should be a true test of what will happen once this fix is moved in with the rest of the live code. Hopefully nothing goes wrong.

Any time new code is put into production, the test team will verify that it's working correctly. I've learned over the years not to call it *testing* once we get into production—this makes people freak out because they think it means that the tool hasn't passed all quality tests. They call it verifying. Anyhow, we not only have to verify that the new code is working correctly, but we also have to perform something called *regression testing*. This is additional testing that the quality team does on the rest of the site(s) to make sure everything else is still working correctly and that the new code hasn't messed anything up. I'll go over this more in the testing section of this book, but it's pretty difficult to find something when you don't know what you're looking for.

So, everything has been deployed into production and our tool looks great. The next thing to look at in Figure 6.3 is how the production environment is copied down to the lower environments. I bet you can guess why. We want all of the environments to have the latest set of code, so that we're testing against a valid backdrop. This requires a little bit of downtime per environment, but it's well worth it to keep everything up-to-date. The developers simply take a break, let the environment get the update, and then drop their new code back in.

Domain Mapping

I just got a call from one of the developers assigned to Jetzen, and they're ready to get started. They just need to know which environment they'll be working in and what the URLs are. I told him that the infrastructure manager has assigned us to development area B, and that I'd get the domain mapping process started. What does this mean? We need to create URLs that will be used to publish our site in each of the environments, including production. We've already established that Jetzen has an existing website—*www.jetzen.com*. Of course, we want to keep that same domain name because customers are already familiar with it and there's a lot of equity built up with SEO and their existing advertising. But, for the new site we don't want to use the exact same domain because it may cause confusion, not only with our internal developers, but potentially, with servers. Remember that as we're building this new website, there's still an existing live version that needs constant maintenance and updating. With that in mind, let's initiate the process of creating the domain mapping and v-host configuration required to publish our site in the testing environments and up into production. We can do something simple with the URL like make it *www.new.jetzen.com* and then when we cutover, we'll just drop the "new" part.

The infrastructure manager is going to handle most of this for us, but we need to tell that person exactly what we're expecting to happen. To get this process started, we should open a ticket for the infrastructure manager to create the virtual host and domain mapping configuration for our build site in development area B, the testing environment, preproduction, production, and the duplicate production environment. Make sure they understand that we're creating a responsive website and that upon cutover we'll re-route to *www.jetzen.com* which already exists. Providing as much information as possible will help them to prepare for final cutover and help ensure nothing goes wrong. Hopefully they were able to make it to the kick-off meeting and this doesn't come as a complete surprise to them.

Just like any other set of code, we need to test our configuration in each environment before moving up the ladder. Our configuration code will probably be scheduled into a release cycle, and will follow the normal protocol for testing and verification until we are approved to be released into production. There's no reason to delay with this, just get the process started so that there's nothing holding us back as we proceed.

With that we've accomplished our goals for infrastructure analysis. How about learning about some fun stuff? Well, fun to me. How does our website make it around the world? Allow me to explain a few definitions which will help us understand. Follow the flow of information around the world in Figure 6.4, while reading these terms:

Virtual Host: At the bottom of the diagram, we see the Jetzen host server receiving content requests and pushing out information globally. The virtual host serves up the Jetzen Spaceship's website, as well as several other websites owned by the same conglomerate that owns Jetzen. This corporation also owns companies who sell things as small as hoverboards, up to larger purchases like the spaceships and even a shipping company.

Domain Mapping: Not shown in the diagram, but this is the process of connecting the domain name (*www.jetzen.com*) with the internet protocol (IP) address of our host server. The user has an IP address too, so when these numbers are exchanged, they can call each other up and start dating. Actually it'd be more like texting!

Content Delivery Network (CDN): This is a system of server nodes located across the globe, strategically placed to provide users with quick response time when they search for content. They locate and publish the correct content when people type in addresses like *www.jetzen.com*. Remember, the content delivery network knows that www.jetzen.com relates to a specific IP address, and they've been provided with a map to that location so they can quickly find and retrieve the information.

These servers also save copies of the sites they publish so that our personal or public networks don't have to go all the way back to the main source of information (host server) every time somebody wants to look at a page. This memory bank is called the *cache*. When the very first person in the world (we'll call him Justin) types in *www.jetzen.com* and looks at our site, every page he views will be kept in the memory of his local CDN server for use with future visitors. This will result in reduced page load times for Justin—and also reduced stress on the host server for Jetzen. Of course, the nodes are smart enough to know if content on the page they are looking for has been updated, and in this case they'd have to go back to retrieve the updated information, but unless there's new content to share, the nodes will just publish what is already in their memory.

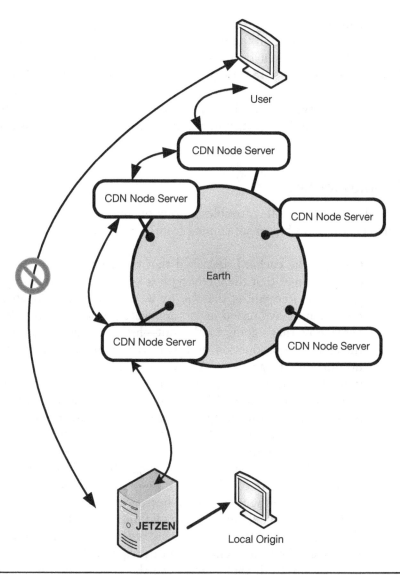

Figure 6.4 When a user *clears their cache* they're ignoring any locally cached files in their computer

CDNs are owned and operated by companies that provide this service to paying customers—so not every website uses something like this. In fact, only huge websites that expect an enormous amount of global traffic to their site purchase this type of service, but it's well worth it.

Let's say Justin is sitting in Frankfurt, Germany and types in *www.jetzen .com*. Instead of calling all the way back to Arizona (Jetzen's headquarters and location of their host server) for the content, Jetzen uses this CDN system of servers to pass the content from one node to another until it reaches Justin. We don't want each view of our website to hit the main host servers—they'd never be able to keep up with the amount of traffic and would, at best, slow things down and, at worst, crash the system.

Troubleshooting

How about learning a few troubleshooting techniques that can be used to detect where problems are coming from?

1. **Clearing the cache:** I described that the cache is a saved bank of page copies that the CDN servers keep stored in their secret stash area. Remember that *hot fix* we had to rush through the process and have tested in the duplicate production area? Well here's how that's going to play out—first we're going to release the code into production, and then we're going to start packing up our stuff to go home. We're done, right? Not so fast—not until the client sees that the change has been made with their own eyes. There we will be, sitting on the phone with the client—we can see the change on our computer, but they can't see the change on theirs yet. What we need to tell them to do is to *clear the cache* on their computer. This directs the server network to re-pull assets from the host server, instead of what it has in memory. In Figure 6.4, I have a no-go line for direct communication between the user and the host server because we're going to set up our global network to avoid this route. However, when there's an immediate need like this, we can use a manual override by clearing the cache. Once this happens the client should be able to refresh their screen, see the change, and everybody can go home.

2. **Check the origin:** What if that hot fix wasn't reaching anybody's computer, even ours? The development team told us that the code was successfully placed into production, but nobody can see the fix yet. One of the first things we do is to check the local origin publisher. The origin site is a version of the website pulled directly from the source, on a special URL that only

internal folks can see. Well, only folks who have the correct user name and password can see it, anyway. Check the origin—if the fix does *not* appear on this screen, then there's something going wrong somewhere at the host server level. But, if the fix *does* appear on this screen, then there's something going wrong with the content distribution network.

3. **Check the publisher cache:** In a case where the fix is not showing up on our origin publisher, but the developer is swearing to us that the code is in there, ask for the publisher cache to be cleared. This is something that the infrastructure team has to help with. Once it's cleared, then the fix will probably show up—but if it doesn't, that means we need to keep looking for the problem.

4. **Check the CDN cache:** In a case where the fix is showing up on the origin, but not on the client's computer—even after they've cleared their cache, then there might be a hang-up with content distribution. Ask them if we need to clear the CDN cache.

Even when these troubleshooting techniques don't solve the problem, at least we know what isn't causing the problem. These are just a few, and I'm sure you'll learn many more along the path to digital project management adventure land. It's all about tracking the bits.

Well, that's about it for the infrastructure assessment. Now let's move on to an even more boring topic—the IT bill of materials (BOM).

IT Bill of Materials

When I was new to digital, and people started talking about the term *IT BOM*, I thought they were talking about some sort of *IT bomb*. Like something was going to blow up. I'm not kidding—I had no idea what they were talking about, until I did the unthinkable and asked somebody.

Big surprise—there was no bomb! They were talking about the information technology bill of materials. Having a bill of materials is not unique to information technology. Any company that has hardware components or parts will have a list of every single component or part in their possession. I assume that most people reading this book have a computer on their desk at work. Well, chances are that the computer has a barcode and serial number taped to it somewhere. In addition, if there's a separate monitor for that computer we'd probably find a barcode on that. Not to mention the power cord, mouse, phone, etc. All of these barcodes are there to help the tech team keep track of all of their *stuff*—which will be listed in the company's BOM.

Jetzen has one too. They have materials not only to run their offices, but also to build the spaceships. When it comes to our website, we need to check the BOM specifically related to hosting websites. Hopefully, creating this new website will result in increased traffic on their servers, which will increase the load on the system. Also, the new website has more tools and a larger technical solution managing it, so that will need to be carefully considered, as well. Finally, just the process of building a new website means that we're doubling the size of the existing structure. Now we'll have two sites—the live site and the duplicate one we're building. Think of all the extra copy, photos, and videos that will need to be supported.

It's another job for the infrastructure manager, and this step comes right after the infrastructure assessment because these two things go hand-in-hand. Upon reviewing everything, the manager will let us know if additional hardware, servers, memory, or whatnot has to be added to their BOM and purchased to support the system. It's that *whatnot* we have to watch out for.

In this chapter we:

- ✔ Analyzed the analytics recommendations
- ✔ Established a process for obtaining the SEO recommendations
- ✔ Made sure the environments were ready
- ✔ Checked the IT BOM against our requirements

7

DEVELOPMENT

I have three brothers who are quite a bit older than I am. My parents had Jeff, John, and Jerry right in a row, and then about six years later along came their little girl. At the time I'm sure people would have guessed that my parents were going to name me Julie or Jessica or something similar to that to go along with all the *J* names, but no—they named me Nancy—and then actually called me *Tai* (Tay) which was short for my middle name, which was Taylor. They were obviously confused and worn out at this point in their lives—but don't blame them—blame the boys.

My dad worked for EDS as a systems analyst, which means he was one of the first programmers to ever walk the earth. That's what we tell him, anyway. I'm telling this story because I have a special place in my heart for developers. My dad started in the days of punched cards—when a computer program was a set of cards with little holes punched all over the place. Throughout the house I'd find stacks of these little cards held together by rubber bands, and if the stacks got too large he'd put them in a shoebox. In today's language each set of cards would be a program, the shoebox would be a *folder*, and if there were several shoeboxes sitting inside of mom's cabinet, then that would be a hard drive.

I also remember my youngest brother, Jerry, attending the University of Michigan to earn a degree in computer science. He was learning to program computers using a software language called PL-1, and he and our dad would review code together at the dining room table after dinner. Many nights I'd watch them together and wish I could do it, too—to be part of that togetherness. When I was old enough, Dad bought me a Texas Instruments computer and gave me his old issues of programming magazines so I could copy programs and make things on my computer. It was really fun, but I'm sure the best thing I ever made was a stick figure waving its hand or something silly like that.

Today people don't necessarily need to go to a university to learn programming skills, many learn on their own using information found on the internet and messing around with their hacker friends. Not that going to school is a bad idea; I'm just saying that I've known many successful developers who didn't have a college degree, but did have impressive resumes and portfolios. There are so many different types of programmers and different *languages* they use—I really don't know the difference between each of the languages or how they're different from one another. That's alright, though, because the functional manager in charge of development at our company will let us know which development resources we should assign to each of the tools we're building for Jetzen.

An astute reader may question why development is placed in the planning section of this book, when it's actually more like *doing*. Development *is* doing, but the entire Jetzen budget has been approved by the clients and they're eager to get it going as soon as possible, so we're going to start development right now versus waiting for each task under the planning section to be complete. I find this to be the case for most projects. If for some reason the clients wanted to wait, we would certainly do that and move development down under Construct to Close, but these people have signed the preliminary budget estimate and they're ready to go. Being able to move steps and adjust the process is an example of how flexible it is.

Regarding the budget, remember how we priced out each tool individually so that we could track costs more accurately? One way to make sure this happens within an agency is to open a job number for each of the individual projects. For example, the contact form would have its own job number; the configurator would have its own job number, and so on. When development starts, make sure everybody is billing to the correct job numbers. Let's take another look at our development timeline (see Table 7.1).

Notice that the first three tasks for development start on the same day, which is March 20. That's because the very first sprint—navigation and homepage—is done being styled on March 19. This is the earliest possible date that we can begin development of the basic elements of

Table 7.1 Development Schedule

66	Development	Fri 3/20/15	Mon 7/13/15		
67	CSS	Fri 3/20/15	Thu 5/28/15	17	Jacob
68	HTML Framework	Fri 3/20/15	Thu 6/11/15	17	Sandy
69	Search	Fri 3/20/15	Wed 3/25/15	17	Tim
70	News Stories	Fri 5/1/15	Thu 6/4/15	51	Paul
71	Video Gallery	Fri 4/17/15	Wed 5/6/15	43	Tim
72	Configurator	Fri 5/15/15	Mon 7/13/15	59	Charlie
73	Contact Us Form	Fri 5/15/15	Mon 6/1/15	59	Tim

the site, which are the Cascading Style Sheet (CSS), framework, and search configuration. Also, since we now know the actual names of the people who will be working on these tools, let's go ahead and enter their names into the project plan and add them to the stakeholder list, if they've not already been added. Accurate record keeping helps not only the current project, but future projects as well. If we ever need to refer back to one of these projects in the future, we can simply look up who developed the original tool and hope they still work for the company.

The developers who will be working on the CSS and framework only have a small first portion of the styleguide ready to work with, but they have enough to at least get started. In order for the developers to see what they're working on, though, they'll need some dummy copy and images to drop in. For instance, the front-end developer (Jacob) is able to get the headlines, subheads, and paragraph text put into place, so either he can just make up some headlines to test or we can send him some dummy copy. Either way—it doesn't matter—talk among yourselves. Once the other sections of the styleguide are done, we can get those developers up and running on their assignments as well, until everything's in motion.

COMMUNICATION

Communication with the development team is imperative. The problem is that these guys and gals don't usually go to the status meetings, although they're welcome to attend, and they have little to no direct contact with the team. They'd rather stay at their desks and get their work done instead of sitting through meetings that may not necessarily pertain to their particular assignments. And I don't disagree—we need them to focus on the tasks at hand, while the client-facing team handles external communications. So how do we best communicate with the developers? Well, it's the project manager's (PM's) job not only to communicate complicated technical concepts from the tech team to the nontechnical folks, but also to communicate nontechnical business requirements back to the tech team. Yes, we need to talk through both sides of our mouths. It's in our job descriptions!

Not much besides good old-fashioned work experience will help us prepare for this role. After months or years of dealing with both sides of the technical spectrum, we start to understand their various terminologies and frames of reference (and we become rollout managers). The best approach is to write it all down and then follow up with verbal communication. It doesn't need to be in person, either. People don't like to be disturbed while they're engaged in their assignments at work. I recommend documenting any requirements that the development team needs to be aware of, e-mail it to their attention, and then ask when would be a good time to follow up over the phone or in person. Of course this type of communication is used for little things that come up along the way, not for huge changes to the project requirements. Those require a separate process altogether—called the change management process.

Change Management

I want to discuss how to effectively manage changes that come up after the scope and budget have been approved. It's very normal, and even advantageous, for the creative or brand teams to come up with additional features they want implemented on the project because, hopefully, those new features will enhance the customer's experience, lead generation, or the sales process. Those are all very worthy endeavors, and we don't want to get in the way of building the best possible product. As the team lead, the PM should create an atmosphere where creative minds are allowed to flourish—it's best for everybody involved. The client will be happy that the team is able to not only come up with such great ideas, but to also get them implemented in a timely manner. We all know how quickly technology changes, so the process needs to be flexible and able to accommodate change. This section is not about obstructing new ideas; it's about controlling and managing them.

Since we've done such a good job at documenting the business and functionality requirements along with their related budgets, keeping changes under control shouldn't be a problem. But change can be very objectionable for PMs. I can admit to having had a very hard time with change in the past. So much so, that I thought about it for a long time and tried to figure out what was driving me to hate it so much. I knew that change was good, but it made me feel angry. I was able to come up with an association between the feelings generated by project change and some feelings I had as a child. I can remember once building a fort out of big blocks in kindergarten only to have the class meanies come through and destroy it. That was so mean and I was so mad! And that's how change sometimes feels to a PM. We've invested hours researching and documenting, building the business requirements, creating and co-ordinating the schedule, and getting all our ducks in a row and headed toward the pond, only to have somebody come through and destroy it with a great idea.

I had to figure out how to get my head in a different place—to try to be more like the creative team. Creative people *love* change. They thrive on it, and it makes them feel like their big block fort is getting bigger and better, and dominating the world with its greatness. Think about it—their whole job is to think new thoughts and create new things. Our job is to foster and implement that creativity. I found that the simple act of having a documented change control process in place pretty much eliminated the stress I felt, because I knew there was a plan in place to handle that change and that it couldn't blindside me.

Change Control Process

Changes need to be assessed and brought into a documented change control process any time after the initiation stage, since we have a base set of requirements confirmed at that point. The change control process becomes exceedingly important after scope freeze has been declared (after the styleguide is complete), since the team should be done writing specifications at this time. In fact, after scope freeze the information architect (IA) resources have probably moved on to other projects and the production team has taken the helm. New requirements can increase resource time and cause the budget to increase, so we need to closely monitor anything new coming in. Here are a few other reasons to get a documented process in place:

- Change requests might require anywhere from a very low to very high effort to implement
- Several low-effort changes will quickly add up to a large effort
- Change requests are unpredictable and can arise at any point
- It's impossible to predict what type of change requests will come in
- It's impossible to predict how many change requests will come in

We may as well embrace change because it's certainly not going anywhere, right? It's the nature of our business and we should support new creative ideas—because they're actually how we make money! The process can be whatever the team decides is the best route to take, but Figure 7.1 shows an example of a process I like to follow.

Figure 7.1 has three swim lanes, one for the PM, one for the core team, and one for the client. Both the PM and the client are also part of the core team, but they have specific roles during the change control process, so I've called them out separately. The core team could consist of all other team members who are part of the weekly status meetings and keep a close eye on project activity, or the client could decide they want a smaller group reviewing and ruling on change requests. Reviewing this process with the team will prove that there's a process in place to assess and decide on change requests so that nobody feels decisions are being made in a vacuum by partisan parties. Here's a step-by-step guide to go along with Figure 7.1:

- **Change request identified:** Change can come in from any team member and via multiple types of communication paths. Make sure the idea is fully understood, and follow up with the requesting party to ensure we scope out the right thing. I can't even begin to explain how many times I've seen people put together a scope based on what they *think* other people are asking for, without going back to ask questions. And this is even when the internal team is clearly communicating that they don't understand what they're supposed to be scoping out! This normally happens when the request comes in from a client. I've never understood why they refuse to ask questions—it's like they don't want the client to know that they don't understand. Asking questions actually proves we *do* understand because we're getting down to the details.
- **Update change control log:** The change control log (refer to Table 7.2) is updated with the name of the requestor and a description

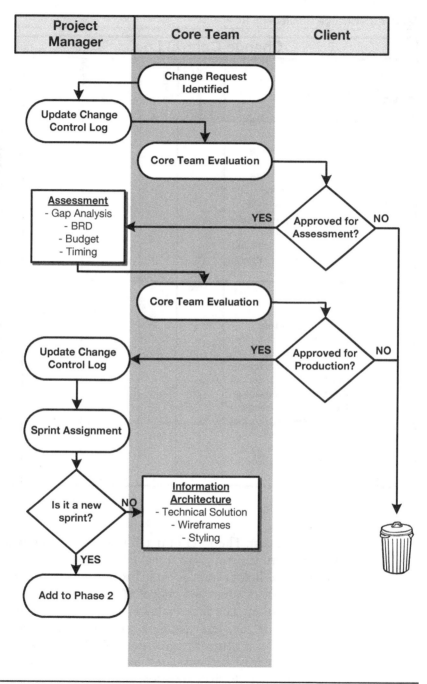

Figure 7.1 Change requests are most likely to be brought forth during testing

Table 7.2   Change Control Log

Change Control Log							
Change Request	Description	Requestor	Evaluation Date	Budget Impact	Go / No Go	Phase	Sprint
Add model maker to configurator	Customers will be sent a model replica of the spaceship they build online through the configurator. Requires customer login with save and send features. Additional data capture and thoughput to third party.	Lindsay	3/11	85,000.00	No		
Trip Planner	Tool to help space travelers prepare for their adventure. Based on travel time and estimated number of miles, the tool will predict gas usage in addition to providing tips for passengers. Tips may include spacesuit instruction, meal planning, anti-gravity techniques, etc...	Pat	3/18	40,000.00	Go	1	5

of the change request. The level of detail put into the log is up to the PM or team. It's really just a log, not a source document, so maybe there's a link or path provided to retrieve the backup information on specifications and budget.

- **Core team evaluation:** The first evaluation the team does is to simply determine if the new change is even worth evaluating further. Does the team generally agree that the change is a good idea and should at least be considered for feasibility? Does the team have an instinct of what the effort level will be to get this change implemented? The change could be rejected for a number of reasons—too expensive, too difficult, too time consuming, etc. But if somebody submits a change request, it's professional and courteous to respond back to that person with the result of the evaluation and the reasoning behind why we are either moving forward with it or not.

- **Approval for assessment:** This is in the client's swim lane because it's ultimately their decision. Of course they'll take input from the core team they've selected to participate in the evaluation, but the budget and timing impact that the change may cause is their responsibility. If the client decides to reject the change request, then the final result is documented in the change log and the idea is terminated.

- **Assessment:** If the client decides that the change is a good idea and that they would like more information, we put it through the normal process that we follow for all other requirements. Note that we may need to get back on the schedule for any resources required for the assessment, if they've moved on to other projects. A gap analysis is completed first—basically determining what we'd need to do to implement the change, and then basic business requirements are documented so the production team has enough information to provide their estimates. The budget is created and run through internal management for approval, and the schedule is assessed.

- **Core team evaluation:** Remember the spaceship model builder idea that someone came up with (let's call this person Lindsay)? The one that built little plastic models using data captured from the customer's configurator creation? Using this as an example, the team assessed that it could be done, but the incremental costs would be $80,000 and it would also extend the timeline. Figure 7.2 is a graphic representation of our development resources and their assignments. Since this new tool is part of the configurator, we'd have the same guy, Charlie, work on the model builder tool—since he's developing the rest of it. The bad news is that the configurator is already the last thing to get done in our production schedule, so adding more to Charlie's job will extend our overall project plan.

- **Production approval:** Due to budget limitations, the client has decided not to move forward with Lindsay's model-builder idea. But just for kicks, let's do another example. One of the second-tier

	March	April	May	June	July
CSS	Jacob				
HTML Framework	Sandy				
Search	Tim				
News Stories		Paul			
Video Gallery		Tim			
Configurator			Charlie		
Contact Form			Tim		
Model Builder					

Figure 7.2   The Model Builder functionality would extend our production schedule if approved

clients at Jetzen (Pat) came up with an idea for a Trip Planner tool. This new page would be designed to help space travelers prepare for their adventures. It would be both fun and interactive with *tips*, videos, and a fuel calculator. For instance, based on travel time and estimated number of miles, the tool will predict gas usage—and video tips can include anything from spacesuit instruction, activity ideas, or techniques for cooking in zero gravity (Heaven's Rider comes with a kitchenette). The team put this idea through internal assessment and the clients just love it. The cost fits within their budget ($40,000) and based on Figure 7.3, we can get it done during our normal production schedule. Notice from the chart that although this is a brand-new functionality, Paul is available to work on it after he gets done with News Stories. Also, he'll get done with Trip Planner about the same time that Configurator is done, so all-in-all, no additional time needs to be added to the schedule. Thus, this change is approved for production!

- **Update change control log:** Next, the log is updated (refer back to Table 7.2) and the PM will send an e-mail to both Pat and Lindsay explaining to them the outcome of our evaluations. Lindsay was bummed that her project didn't make the cut, but of course, Pat was thrilled.
- **Sprint assignment:** We have to refer back to the schedule to figure out which sprint Trip Planner will fit into. Trip Planner was officially approved by the evaluation team on March 28; this was

	March	April	May	June	July
CSS	Jacob				
HTML Framework	Sandy				
Search	Tim				
News Stories			Paul		
Video Gallery		Tim			
Configurator				Charlie	
Contact Form				Tim	
Trip Planner					Paul

Figure 7.3 Paul is going to develop the new Trip Planner tools after he finishes News Stories

before scope freeze but after most of our sprints had already been wireframed and styled. We're left with the last sprint—number five. Notice from Table 7.3 that I've inserted all developer names and added the Trip Planner task. If there were no sprints available to put this into, we'd have to consider adding this new functionality into Phase 2 of the project. We use the throw-all *Phase 2* language to represent things we don't have to do right now—we'll do them later. Or, the sustain team will do them after we move on to another project.

- **Information Architecture:** The first thing the IA will do is to update the sitemap with this new page. Even though the sitemap has already been officially approved, we'll send it out once more with

Table 7.3 Development Schedule with Trip Planner

66	Development	82 days	Fri 3/20/15	Mon 7/13/15		
67	CSS	50 days	Fri 3/20/15	Thu 5/28/15	17	Jacob
68	HTML Framework	60 days	Fri 3/20/15	Thu 6/11/15	17	Sandy
69	Search	4 days	Fri 3/20/15	Wed 3/25/15	17	Tim
70	News Stories	25 days	Fri 5/1/15	Thu 6/4/15	51	Paul
71	Video Gallery	14 days	Fri 4/17/15	Wed 5/6/15	43	Tim
72	Configurator	42 days	Fri 5/15/15	Mon 7/13/15	59	Charlie
73	Trip Planner	26 days	Fri 6/5/15	Fri 7/10/15	71	Paul
74	Contact Us Form	12 days	Fri 5/15/15	Mon 6/1/15	59	Tim

this new revision—updating the version number also. As shown in Figure 7.4, we've placed this page under Owner Resources—too bad we weren't able to incorporate Trip Planner into the Owner Resources sprint, but it's already done.

Agile Methodology

There are some things in this chapter that would actually never be allowed in a true agile work environment, so call me a rule breaker. The first one I'll mention is allowing the new requirement to fit into the current phase, in one of our current sprints. Not everybody's going to agree with me on this—and please make sure to follow whichever process your agency's management wants to implement—but we're in the business of customer satisfaction, and that means the customer comes first. This refers to both our customers and Jetzen's customers. We're also working in a very competitive industry, and anything we can do to remain organized, yet flexible, will help our clients to succeed and, in turn, reflect positively on our agency.

From my understanding, the agile methodology prohibits any additions to sprints. This is where I feel we can be a bit less rigorous than what the books tell us to do. Why not be open to it? It's the client's money and schedule—if they don't mind spending a little more money or time then why wouldn't we accommodate them? From the agency's perspective, we've put a cost against it and have determined that we do have the resources to fulfill the request, so it's really just good business to accommodate the customer. Now, if we've decided that the new request couldn't be done under the same timing or that we'd have to hire more resources in order to get it done, then our job is to present the clients with the facts of the case and let them decide how they want to proceed. It's their project, really.

Another small thing I don't put in my process are the daily stand-ups that agile methodology recommends. May I just say, "No, thank you"? There's no reason *not* to meet face-to-face, but I'm just not going to be

Figure 7.4   Trip Planning is a natural fit for the Owner Resources group

the one who puts this meeting on everybody's schedule for first thing in the morning. If you need to meet, just meet. It doesn't have to be mandatory. Maybe I just don't like standing up, or meetings, or mornings, etc. There are actually a lot of things about my process which probably don't align with true agile methodology, but every project is different and should be looked at with a fresh perspective on how to manage it.

Development is the fun part. Once these people get going, we can start to see our vision come to life—ideas become reality. Have fun with it!

SOMETHING EXTRA—DATA PLANNING

When the team is creating forms or other applications, there's usually a pretty strong data component that needs to be rigidly composed and delivered to the development team. For instance, for the contact form it might look pretty easy for a developer to put this thing together, but they'll have a lot of questions that we should try to predict and document so they don't have to research the answers themselves. Table 7.4 shows just a few of the questions we should have documented for every single entry point on the form. This includes every line—name, phone, preferred contact method, inquiry type, and even the comment section. Let me go through them and explain—just in case they aren't obvious:

- **Visible form element:** Forms or applications can grab information from a customer contact without the customer inputting the information into the form. For instance, the form could be created to pick up the customer's internet protocol (IP) address or to attach a reference number to each submitted form for tracking purposes. These are things that will be sent through with the data, but the user never initiates them and there are no related fields on the form. This brings up another interesting point—how is the data actually recorded and received? The internet uses a common language for computers to talk to each other called Extensible Markup Language (xml) syntax. When a customer submits a form, the xml file will capture all the text they type in, the selections they make, their comments and nonvisible elements that it's supposed to capture. When the data is received, the developers can pull the xml file and review the captured elements just to make sure everything is coming through as expected. Otherwise, the data will just be sent through to the appropriate web services or databases.

Table 7.4   Data Mapping Example

Visible form element?	Y
Mandatory or optional?	M
Field type	Alphabetic
Maximum field length	55
Minimum field length	9

- **Mandatory or optional:** Sometimes companies can get pretty aggressive with the amount of information they want to get from their customers. There can be so many fields to fill out that customers get frustrated and close out the form because they either don't have time to fill everything out or they feel their privacy is being compromised. Because so many people were dropping out, we started implementing mandatory fields, which are usually marked with an asterisk, to indicate which fields the customer absolutely needs to fill out in order to complete a submission. For example, if the customer is asking for an e-mail reply from Jetzen, then we definitely need their e-mail address to be a mandatory field. This way the customer decides if they want to submit the *extra* information—or if they just want to fill out the minimum requirements.
- **Field type:** Some examples of field types are alpha (meaning the field will only contain letters), numeric (meaning the field will only contain numbers), alphanumeric (the field will contain both numbers and letters), or a coded value. The first several are very easy to understand, but what do we mean by coded value? Take the example of Jetzen's contact form as shown in Figure 7.5. Notice that the customer is asked to select their inquiry type. The customer will use the drop down menu to make a selection and the xml will compare that selection against a list of available types and their associated codes. Then the xml will grab the right code and pass it through with the rest of the data. Of course, the code could be either alpha or numeric (or a combination of both), but the fact that we list this as a *coded value* means that there's a list of possible entries and their associated codes. We may be required to use these codes because the database receiving our xml will only understand what to do if we use the right codes.
- **Maximum and minimum field length:** This indicates the number of key characters the field should be set up to accept. This can be quite helpful in cases where validation is required and could result in an error message, if not followed. For instance, for a phone number we would want to make sure that the customer inputs ten digits exactly (i.e., 248-555-5555 not counting hyphens). If they input any less or more

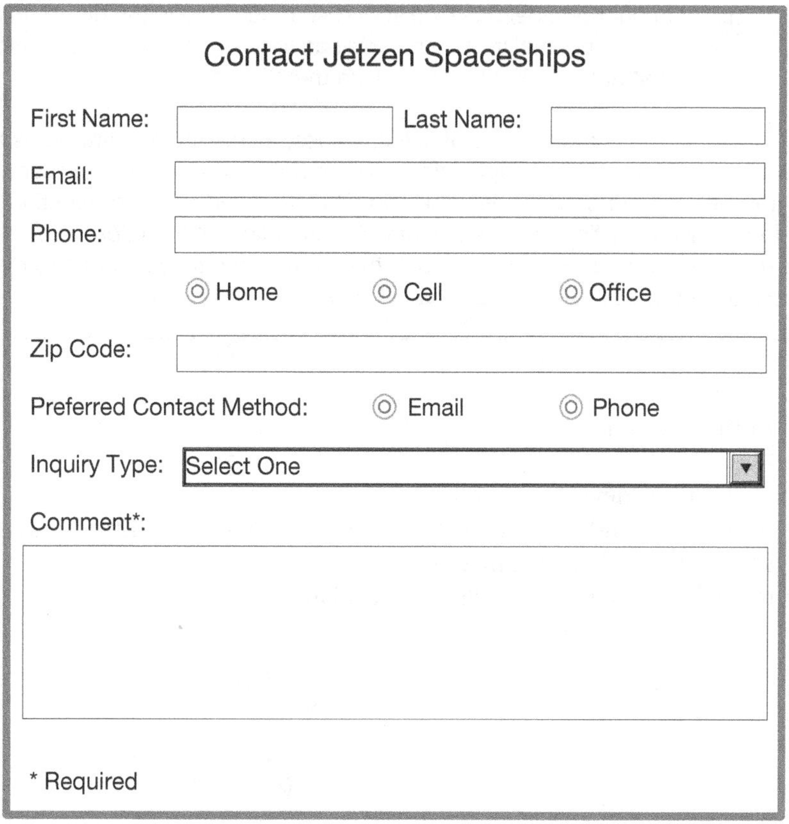

Figure 7.5 Reviewing the Jetzen contact form should help to understand the terminology described in this section

than ten digits, then we're not going to get a valid phone number, and if a phone number is a mandatory field, then we need to make sure ten digits are captured. Now if this were a global form, we'd have to add more digits to that field to allow for country codes.

- **Validations and error messages:** Related to all of these errors we've been discussing are the validations and error messages that need to be documented and tested. As I explained using the phone number example, we need to be really careful about what counts as an invalid entry and what the appropriate error messages might be. In this case, any telephone entry with more or less than ten digits would result in an error message that said something like, "Telephone field requires ten digits." Most of us should be experienced with error messages at this point in our lives, because we spend so much time in the digital space even in our personal lives, but a PM has to remember that these things don't just happen. We need to make them happen.

There are many more types of fields, values, elements, and attributes associated with forms and applications. The point here is not to list them all, but to understand how things work and what we need to do to help the developers use their time most efficiently. By providing all the information they need to work on our projects up front, they don't have to wait for us to mess around with what should have already been done.

In this chapter we:

✔ Got our developers up and running
✔ Created and communicated the change management process
✔ Started a change control log
✔ Added in a new tool called Trip Planner

8

PREPARATION

We've really accomplished a lot since getting that first phone call, haven't we? Things are really moving now—the creative team has started production and our developers have started programming. There are just a few more things we need to tidy up before Plan and Define is officially over and we're ready to enter the Construct to Close stage. The first thing we'll discuss in this chapter is testing. This is an area that's pretty easy for a project manager (PM) to overlook since most large agencies have a test (or quality assurance) team in place to manage this part without us getting too heavily involved. But please be sure not to ignore it altogether, because things can fall apart pretty quickly if we don't make sure the test team is fully briefed and prepared for the project to be handed over to them. And I speak from experience. Before I fully understood the process that I'm explaining in this book (remember, there wasn't any documentation for how to do my job when I started), I was in a situation where my digital project was ready to be tested, so I called up my test lead and asked if they were ready to go. I found them to be fairly confused. "What is it supposed to do? What are we testing against?" Those are both good questions. I didn't know that I was supposed to have asked for a test strategy and test cases ahead of time—thanks, nobody. So, to help ensure that no one else is faced with a blank stare when turning over a project to the test team, I thought I'd go ahead and document the process. The first thing we need is a test strategy (some companies call this a test plan).

TEST STRATEGY

The *test strategy* is a document that describes the testing standards established for both the agency and the project we're working on. It describes how the testing will be conducted, what will be included during the test cycle (and what won't be included), the criteria that'll be used to determine if the test cycle has been successfully completed, the method for reporting defects to the client, and the proposed schedule. This document doesn't have to be updated from scratch every time we need one because many of the subjects described in the document don't change—they're standards our agency follows for every project. But we do want to tailor the document for this project and deliver it to the client for approval, especially in the case of Jetzen, since we've never dealt with them before. They've never had an opportunity to review our test strategy document in the past, so they're probably eager to read through it (Ha! That's funny to me—*eager* to *read through* a test strategy...). Some of this information may have been presented to them during the initiation phase, or certainly, if we had been responding to a formal request for proposal (RFP), because a company that issues an RFP would probably ensure that the agency they're considering for the project has substantial testing and quality processes in place.

This *may* not be the most interesting part of this book (it's not), but I want to go through what a test strategy document normally contains, because the terminology will definitely come up—if not during our test cycle, then for sure at some point during one's career. I want to provide the reader with reference material in case one of these terms comes up, but isn't fully explained. There are also a lot of good resources on the web, so if you ever come across a term that can't be found in this book, just search for it online.

Inputs to the test strategy are the business requirements document (BRD), wireframes, and styleguide. The Jetzen test team knew ahead of time that they were going to be asked to create a test strategy, because I told them about it back when we were talking about the original budget and confirming resource availability. But it's something that can't start until the three inputs become available, so they may need a refresh. Figure 8.1 shows what the table of contents for a strategy document may look like. Take a quick look, then I'll go through what each of the sections is all about. Notice that I put this example together myself, based on my own experience. I'm sure that a qualified quality practitioner at your company will have a different template to follow.

Contents

Figure 8.1 A test strategy should explain the agency's theory and practice around quality assurance

Scope and objectives: This could mean a couple of different things. It may include the scope and objectives of the test strategy document itself, or the scope and objectives of the project.

Business requirements: This is a highly customized section of the document since it should describe the business requirements of this particular project. Notice that we're not really talking about functional requirements at this point—those will come with the test cases. Right now we just want to communicate that our test team understands the scope of the Jetzen project and what they'll be required to test.

Roles and responsibilities: This part isn't always necessary, but the team may elect to document which resources are scheduled to work on Jetzen. We could mention who the quality assurance (QA) lead will be, as well as the names of the testers, but it's a touchy subject to mention actual names of resources in any document, since business changes so frequently. These people could move to another project or another agency by the time testing actually starts, so I don't recommend naming names. For Jetzen, I'll make sure we state the number and types of resources required for this project.

Test deliverables: You didn't know there could be deliverables associated with testing? Well, there are at least a couple of different reports that the test team may wish to distribute. Testing produces a lot of data—it's all quantitative—so the team is going to produce a weekly report showing the number of test cases performed, the number of defects found, and the severity level of those defects. Also, at the end of the project perhaps, the test team delivers a summary report showing overall activity and the status of the project upon final cutover.

Standards: How does our agency determine if a project is ready for release? What types of standards do we uphold? This is where we provide a list of tasks that the quality team is going to perform, in order to confirm quality. This list would include things like: checking for broken links, checking image weights and sizes, confirming end-to-end testing is successful, checking all content against approved creative, or testing for browser compatibility. It could also be a place to confirm what our agency will *not* be checking for—such as content legal compliance, spelling and grammar errors, or the quality of the content itself (does the copy we are inserting make sense?). Since our agency is just in charge of production, we don't have any responsibility to check what the creative team put together. It's not that we don't *want* to do those things (honestly we really don't), it's just not our job—we're not qualified to judge the creative.

Devices and browsers: In this section we'll list the types of devices and browsers on which the Jetzen website will be tested. We'll go ahead and list the actual types of phones (like Android or iPhone) and tablets that the testers will use to test the website. Since this is a responsive website, we'll want to make sure a variety of screen sizes are used to confirm that the site is presented well on any size of viewport. We also want to test it in several different types of browsers, such as Firefox and Chrome, to ensure that every user, no matter how they arrive at our site, will enjoy the same experience. It's important to list the versions of each of those devices and browsers as well. For instance, as I write this book, my Firefox browser is using 36.0.1. It can be very difficult to keep up with the different models, brands, and versions of all these devices and browsers.

Automation: I was once launching the exact same website in 21 different countries. There was very little difference between the 21 sites—they varied in just the URL name and some slight product offering differences. We also had about five tools in each of these websites which were *exactly* the same, so our test team decided to write an automated test script to save time and money. That was pretty impressive. Instead of an individual having to go through each of the five tools on each of the 21 websites, the test team built a script that did it automatically. It was so cool—like a ghost was moving around the screen clicking on all the right buttons, checking to make sure the text was correct, and actually writing in answers to the form questions. That script really saved us an enormous amount of time. Although we don't have this situation with Jetzen, automated testing could be valuable for on-going regression testing, which we'll learn about shortly.

Defect reporting and tracking: There are various systems that testers can use to log and track defects. We just want to let the clients know which system we'll be using, and why we selected this particular tool. It's also a good idea to show them any dashboard tools the system has in place to monitor the defects.

Test process: This is a heading used to describe a number of functional tests which will be done for our project: unit, systems integration testing (SIT), end-to-end testing, user acceptance testing (UAT), and regression testing. Following this paragraph, each of these are listed and explained.

Unit testing: Also referred to as component testing, this is a test for our individual requirements or tools. It describes how all of our tools, like News Stories or the Contact Form, will be tested individually in their development areas before being released for formal testing by the quality team. Remember how we provided the developers with an environment and URL they could use to publish their work? They were provided for the unit testing, so the developers could make sure that their assignments were working to the best of their knowledge, before moving into the formal testing process. Because the data isn't yet being imported and filtered into our system, the developers use *dummy* data to try and make sure the tools will work as expected.

SIT: Systems integration testing is the start of the formal testing process and is also performed in one of the lower environments, such as our testing environment. All of the individual tools are now incorporated together with the rest of the website, so we can make sure they're all working together as expected. That's why it's called *systems integration*—because we are now joining together different systems that have previously only been tested on their own. At this point, all of the data architecture should be in place so we can begin some real tests and watch as our tools start to use real data instead of the dummy data. As an example, for News Stories we should be able to accept the text and images provided by Jetzen's media feed, and all of our filters and other systems should be in place to make this thing work. We should also be able to send *out* data, enabling tools such as the contact form to send data through to Jetzen headquarters.

End-to-end testing: The contact form is a perfect example of why end-to-end testing is important. In this case we were asked to send the data collected on the contact form to the *Receiver X* database located at Jetzen's corporate offices. Part of the data architect's job is to make sure that the connection is made, and then once it is, we can test the connection during end-to-end testing. To conduct this test simply fill out a form in the test environment, press the submit button, and then the people at Receiver X will let us know if the data is received as expected. If not, we'll have to track down the problem.

UAT: User acceptance testing is *kind of a big deal*. As one of our last steps in the testing process, this is when the website will be moved up into preproduction to get tested by all outside parties including the clients. Be open to the idea of making some changes. Not necessarily functional changes, unless there are legal implications that will impact our ability to launch, but content changes should certainly be expected and implemented as much as possible. We may put a provision in here, stating that we'll accept up to a certain percentage of content changes during UAT, and anything beyond that point will result in a change to the launch date. There is a danger in trying to push through *too* many changes right before launch, as we never know what will trigger a defect. In the last week or so of production, stability of the site should be a higher priority than updating the copy.

Regression testing: I've described this process before, but just to reiterate, regression testing takes place after we move the site into the production environment. If there are other live sites held within this same environment as where we're putting Jetzen, then those sites, even if their code hasn't been touched, will need to be tested just to make sure they're still functioning properly. Any time new code is introduced into an environment, there's a possibility that it could adversely affect other code. In addition, once the sustain team takes over they'll be updating the site constantly, and any code-related updates will require that everything else in the environment be regression tested. It sounds pretty tedious, and it is, but the quality team has approved test scripts in place that they'll follow to complete this work systematically. The team may also employ automated test scripts for regression testing to save themselves from the tedium and long hours of this task. Regression testing is very time consuming but vital to ensure the stability of our projects and environments.

Nonfunctional testing: This is a category of testing which describes behind-the-scenes systems and processes put in place in case of emergency or for security reasons. If our agency were hosting Jetzen's system, then we'd be responsible for all the nonfunctional testing. However, since Jetzen is hosting their own system, we'll just be providing input and support for the majority of the nonfunctional tests. It's important to conduct these tests against a finished product, so we need to schedule the testing to take place parallel to UAT and in a similar preproduction type environment. Since all coding should be complete prior to UAT, and since no structural or architecture changes are expected during UAT, it's the perfect time for nonfunctional testing to take place. It's also important to make sure the environment we use for this testing is as close to production as possible in order to mimic a real scenario.

Performance testing: This set of tests measures our finished product against compliance standards put in place either by our client or by our own team. These tests are extensive and designed not only to measure things like page load times and responsiveness, but also to test the stability of the site and its environment under harsh conditions. Stress testing and spike testing are invaluable to ensure our website is ready for heavy activity and that it won't go down under pressure.

Security testing: When talking about security there are two things to be concerned about. First, we need to ensure that nobody can hack into the site and cause any harm or alter it in any way. We've all heard about hackers breaking into systems and bringing them down or publishing their own messages over another company's domain. This type of security testing is done by the developing party (our agency) and is usually conducted with software that scans the code looking for security vulnerabilities.

Another key thing to consider is protecting our customer's confidential information and data. To show how this works in action—go to a desktop computer or tablet and bring up any major website—what I'm about to describe isn't always clear on a mobile device. Find their contact form (most companies have them) and as you click on that page notice how the URL address changes. A domain that used to start with *http://www* (the *http* part probably doesn't show, but know that it's there in the background) now probably starts with *https*. That little *s* indicates that the user is now on a secure URL and that the personal information they are about to send will be safe and not accessible to any other party other than the intended recipients. On a phone this security feature may be represented by a little lock icon.

Disaster recovery: This is exactly what it sounds like it might be. Policies, procedures, and systems put in place to recover vital technologies in the event of a natural or man-made disaster. Nobody likes to think about it, but what if a tsunami came through and wiped out Jetzen's global headquarters? Well, that would be pretty amazing since Jetzen's headquarters is in Arizona and about 200 miles from the ocean, but something could happen. And if it does, there'd better be a back-up version of that website somewhere, right?

Failover: This is important when the website or system is required to be available 24/7 or 99.99% of the time. That means the client will absolutely not tolerate any downtime and probably has major fines written into the contract if it happens, so we need to have a failover system in place. With the no-downtime requirement it's necessary to have redundant hardware and servers in place and to test them to confirm that if one fails the other will automatically take over and handle all the traffic.

In summary, the test strategy document mostly lists the things that we *will* be doing to ensure a quality deliverable, but it also lists what we *won't* be doing. Commitment goes both ways and we want to be clear about our responsibilities so there's no hard feelings at any time during the process. Once the test strategy document is ready, please send it to the client for review. More than likely they'll ask what they're supposed to do with it—and we can just tell them to make sure they agree with everything it says and send us an approval. Once it's approved we can have the test team start on the test cases for our project.

TEST CASES

Test cases (also called test case scenarios) are step-by-step scripts written for the test team so they have explicit directions to follow, in order to ensure accurate and articulate test results. As I was saying before, if we were to just hand over one of our tools to be tested, the team would be very confused about what they should be testing for. They don't know the project inside and out like we do. And, it's actually a best practice for the tools to be tested by a different team than the one that created it—so it's a good thing that they don't know the tool as well as we do. They're not supposed to. As always, let's look at an example. Test scripts should start out with the project name and some pertinent project information as shown in Figure 8.2. This is an example of a test script written for the contact form. The form can contain a number of different elements, but we definitely want our testers to list their name, the date the test was performed, what was tested (test name), and the testing environment.

Table 8.1 shows how the testing instructions give the tester specific instructions for what to do, leaving nothing to chance. Step 1, for example, has the tester opening the homepage for www.jetzen.com and just confirming that the homepage does indeed appear.

Most test case templates provide a description of the expected result along with a comment area for the tester to explain what the actual

Project: Jetzen Spaceships			
Test Date:	6/3/2015	Test Name:	Contact Form General
Tester Name:	Danielle Howcroft	Environment:	Testing Environment A

Figure 8.2 The test script should contain pertinent project information

Table 8.1 Example test script

Step	Test Step
1	Go to www.jetzen.com
2	Scroll to bottom of page and select Contact Us
3	Check all fields against provided comp
4	Enter data for all fields
5	Select a radio button for Phone type

result was (see Table 8.2). In addition, there's a place to list if the system *passed* or *failed* each step. The example we're following shows a very small list of steps for the contact form just for our example—in reality the list of steps would be much longer than what is shown here. Also, there will probably be several different scripts for the same tool. Another script we'd probably create would be to ensure that the appropriate error messages appear on the screen if the user makes a mistake or omits required information. This list of testing steps would lead the tester to purposely make an error so that the error messages would appear and therefore could be recorded and confirmed.

It's not necessary to send these test scripts to the clients unless they ask for a review. I never list test cases as a deliverable, but they might be written into the contract as belonging to the client since they are paying for the work, so be prepared to turn them over upon request. I don't normally hand them over because it's really the agency's job to ensure that the project is tested correctly. But please do go through the scripts and make sure that the person who put them together didn't miss anything. Remember to refer back to the BRD for information on the expected functionality, and also look at comps to make sure that no additional functionality has been added without being captured in the BRD.

PACKAGE IDENTIFICATION

This section should be pretty short. All we need to do with package identification is to determine which assets the creative team will be sending over on which dates. With this information we can finish out the rest of our schedule. I realize that back in Chapter 4, I've already

Table 8.2   More of the test script

Step	Expected Result	Actual Result	Pass / Fail	Notes
1	Homepage appears			
2	Contact form appears			
	Secure https is visible			
3	Fields appear as shown in comp			
4	All fields accept text			
5	Center of selected button turns gray			
	More than one button cannot be selected			

explained how to finish out the schedule using the package information, but I want to remind the reader that the schedule doesn't actually get completed until the end of Plan and Define. Remember, I said that we'd be updating the schedule throughout our activities and each step builds on the last? One reason it takes so long to get the packages identified and delivery dates posted is because the creative team can't really determine this until the styling is complete and their comps are approved. It's another situation where there's just no use scheduling a resource too far in advance because our business changes so often and by so much that forecasts are often difficult to achieve. At any rate, the package information should be one of the last pieces we need to complete the schedule.

Here's a reminder of the asset delivery dates we came up with back in Chapter 4 (see Table 8.3). Tip: Don't let the creative team provide just one date by which they'll deliver all of the content. First of all, it's too much to go through all at once. Second, even though they'll promise to deliver sets of content as they're approved, don't believe it. We'll end up getting everything at 11:00 pm on the due date. And, I'm not being a jerk here; I'd do the same thing. I've been in their position and the content just never stops changing.

I picked three packages just to make the schedule for this book easier to create. In reality there'd be more than three packages and they'd probably more closely resemble the number of sprint groups. Let's review those five groups once more:

1. Navigation and Homepage
2. About Jetzen and Owner Resources
3. Choosing a Spaceship
4. Jetzen News
5. Help and Tools

Here's a fun exercise—take a look at Jetzen's homepage wireframe in Figure 8.3 and try to figure out which exact assets we'll need to build

Table 8.3 Asset delivery dates

75	Asset Delivery	16 days	Mon 6/1/15	Mon 6/22/15
76	Group 1	1 day	Mon 6/1/15	Mon 6/1/15
77	Group 2	1 day	Mon 6/8/15	Mon 6/8/15
78	Group 3	1 day	Mon 6/22/15	Mon 6/22/15

Homepage

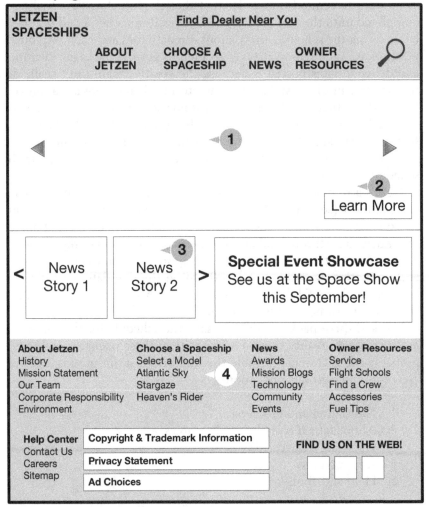

Figure 8.3   Review the page wireframes or comps to determine which assets should be expected for each page

that page. We'll use this information to build our Content Tracker in the next chapter. See, this is all coming together. Trust me.

The assets needed to build that page are:

1. Meta data for search engine optimization (SEO)
2. Desired URL
3. Jetzen logo (All images should include alt tags, naming conventions, URL links if clickable, and copy if not baked into the image)
4. Microscope image
5. Arrow images (unless the developer is able to make them based on the styleguide)
6. Learn More buttons (unless the developer is able to make them based on the styleguide)
7. Masthead images and any associated copy
8. All News Stories teaser images and associated copy
9. Special Event Showcase image and copy
10. Social icons with copy and links showing where they're supposed to connect to
11. Sitemap copy

CONSTRUCT TO CLOSE SCHEDULE

We're finally at the end of Plan and Define! Along with finalizing the Construct to Close schedule we want to have a formal wrap-up to Plan and Define. Let's put a bow on this bad boy and get it approved. We'll have binders—slide presentations—conference calls! This is really a big deal—we've done a great job so far and the clients are thrilled.

In Chapter 4, I promised that I'd show my version of a critical path for this project. Figure 8.4 shows the steps that have to be done consecutively, with the ancillary steps happening in tandem along the way. Gap analysis is first, followed by the client signing the statement of work (SOW), then we make a sitemap followed by wireframes, styleguide, development, testing, and finally cutover. I hope it's clear why these steps must be completed in a row with finish-to-start dependencies. If not, find my contact information and let's connect!

Now let's take another look at our Construct to Close schedule. We've had to adjust it a bit since Trip Planner was added in. Plus, now that we know what the weekly environmental release schedule is, we

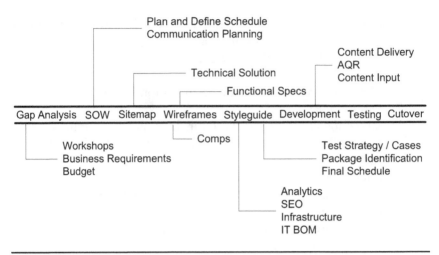

Figure 8.4 Critical path for digital project management

can update the schedule to reflect those dates as well. This release schedule adjustment can be found in Table 8.4 by looking at the predecessor numbers such as Line 105 where we see 74FS + 2 days. I had to put in some extra days because now I know that the pre-environment releases happen every Thursday. Take a look and see if you can follow the logic of the Construct to Close activities.

I showed the final schedule to our client and they had one problem with it. Any idea what would make the client nervous? How about the fact that I only gave them two days for UAT (final team review)? That could be a problem. I knew what I was doing—but this client really wanted to launch the week of August 3, and I wanted to make that happen for them. I'm not expecting the client to give us any more time on the schedule because if they told their boss that the site would launch the week of August 3, then we're launching the week of August 3. So we have to figure something out. Here are a few options we can provide to the team/client:

1. The client can begin their review at the same time that the creative team starts theirs, which will give the client five additional days. This should be enough time for the client review, but the creative team's not going to like it. They'll want to run through the site first to make sure they didn't miss anything.

Table 8.4 Construct to Close Schedule

66	Development	82 days	Fri 3/20/15	Mon 7/13/15		
67	CSS	50 days	Fri 3/20/15	Thu 5/28/15	17	FE Dev
68	HTML	60 days	Fri 3/20/15	Thu 6/11/15	17	Web Dev
69	Search	4 days	Fri 3/20/15	Wed 3/25/15	17	Dev 1
70	News Stories	25 days	Fri 5/1/15	Thu 6/4/15	51	Dev 2
71	Video Gallery	14 days	Fri 4/17/15	Wed 5/6/15	43	Dev 1
72	Configurator	42 days	Fri 5/15/15	Mon 7/13/15	59	Dev 3
73	Trip Planner	26 days	Fri 6/5/15	Fri 7/10/15	71	Paul
74	Contact Us	12 days	Fri 5/15/15	Mon 6/1/15	59	Dev 4
75	**Asset Delivery**	**16 days**	**Mon 6/1/15**	**Mon 6/22/15**		
76	Group 1	1 day	Mon 6/1/15	Mon 6/1/15		Creative
77	Group 2	1 day	Mon 6/8/15	Mon 6/8/15		Creative
78	Group 3	1 day	Mon 6/22/15	Mon 6/22/15		Creative
79	**Internal AQR**	**17 days**	**Tue 6/2/15**	**Wed 6/24/15**		
80	Group 1	2 days	Tue 6/2/15	Wed 6/3/15	76	Asset Mgr
81	Group 2	2 days	Tue 6/9/15	Wed 6/10/15	77	Asset Mgr
82	Group 3	2 days	Tue 6/23/15	Wed 6/24/15	78	Asset Mgr
83	**AQR Fixes**	**16 days**	**Thu 6/4/15**	**Thu 6/25/15**		
84	Group 1	1 day	Thu 6/4/15	Thu 6/4/15	80	Creative
85	Group 2	1 day	Thu 6/11/15	Thu 6/11/15	81	Creative
86	Group 3	1 day	Thu 6/25/15	Thu 6/25/15	82	Creative
87	**Content Input**	**20 days**	**Fri 6/5/15**	**Thu 7/2/15**		
88	Group 1	5 days	Fri 6/5/15	Thu 6/11/15	84	Dev 1
89	Group 2	5 days	Fri 6/12/15	Thu 6/18/15	85	Dev 2
90	Group 3	5 days	Fri 6/26/15	Thu 7/2/15	86	Dev 3
91	**Internal QA**	**17 days**	**Fri 6/12/15**	**Mon 7/6/15**		
92	Group 1	2 days	Fri 6/12/15	Mon 6/15/15	88	QA
93	Group 2	2 days	Fri 6/19/15	Mon 6/22/15	89	QA
94	Group 3	2 days	Fri 7/3/15	Mon 7/6/15	90	QA

2. Group 1 of the creative assets is out of internal QA on June 16—we could start the creative review at this time, but it is well before UAT begins. This is actually a good idea because it gives everybody much more time to make corrections if need be.
3. We could invite the clients to review the tools immediately after they're approved by the internal teams during SIT.

I actually use all of these tactics regularly during my projects because I like to provide people as much time as possible to review the project. There's no reason to hide anything—let's figure out if anything isn't

Table 8.4   Continued

95	SIT	84 days	Thu 3/26/15	Tue 7/21/15		
96	Search	5 days	Thu 3/26/15	Wed 4/1/15		
97	Code in Pre-Prod	1 day	Thu 3/26/15	Thu 3/26/15	69	
98	Internal Review	2 days	Fri 3/27/15	Mon 3/30/15	97	
99	Code Fixes	2 days	Tue 3/31/15	Wed 4/1/15	98	
100	Video Gallery	5 days	Thu 5/7/15	Wed 5/13/15		
101	Code in Pre-Prod	1 day	Thu 5/7/15	Thu 5/7/15	71	Tim
102	Internal Review	2 days	Fri 5/8/15	Mon 5/11/15	101	QA
103	Code Fixes	2 days	Tue 5/12/15	Wed 5/13/15	102	Tim
104	Contact Us	5 days	Thu 6/4/15	Wed 6/10/15		
105	Code in Pre-Prod	1 day	Thu 6/4/15	Thu 6/4/15	74FS+2	Tim
106	Internal Review	2 days	Fri 6/5/15	Mon 6/8/15	105	QA
107	Code Fixes	2 days	Tue 6/9/15	Wed 6/10/15	106	Tim
108	News Stories	5 days	Thu 6/11/15	Wed 6/17/15		
109	Code in Pre-Prod	1 day	Thu 6/11/15	Thu 6/11/15	70FS+4	Paul
110	Internal Review	2 days	Fri 6/12/15	Mon 6/15/15	109	QA
111	Code Fixes	2 days	Tue 6/16/15	Wed 6/17/15	110	Paul
112	Trip Planner	4 days	Thu 7/16/15	Tue 7/21/15		
113	Code in Pre-Prod	0 days	Thu 7/16/15	Thu 7/16/15	73FS+3	Paul
114	Internal Review	2 days	Thu 7/16/15	Fri 7/17/15	113	QA
115	Code Fixes	2 days	Mon 7/20/15	Tue 7/21/15	114	Paul
116	Configurator	4 days	Thu 7/16/15	Tue 7/21/15		
117	Code in Pre-Prod	1 day	Thu 7/16/15	Thu 7/16/15	72FS+2	Charlie
118	Internal Review	1 day	Fri 7/17/15	Fri 7/17/15	117	QA
119	Code Fixes	2 days	Mon 7/20/15	Tue 7/21/15	118	Charlie
120	UAT	11 days	Wed 7/22/15	Wed 8/5/15		
121	Creative Review	5 days	Wed 7/22/15	Tue 7/28/15	119	Creative
122	Fixes	3 days	Wed 7/29/15	Fri 7/31/15	121	Dev
123	Client review	2 days	Mon 8/3/15	Tue 8/4/15	122	Team
124	Fixes	1 day	Wed 8/5/15	Wed 8/5/15	123	Dev
125	Cutover	1 day	Thu 8/6/15	Thu 8/6/15	124	Inf

right and get it fixed as soon as possible. I use UAT for a sort of *last chance* final testing period. This may be another example of me *breaking the rule* that external parties are only supposed to review the site during UAT. Who made these rules? [Robot voice] "You must comply...or face imminent death...or dismemberment...you can choose which one you want...death or dismemberment...or to comply...which we strongly recommend."

After my consultation, the client decided to go with Option 1—plus Option 2—plus Option 3. Whoops, sorry! I should have known. So now we need to adjust the schedule so we can let the clients know which dates they need to set aside for this additional testing. But there's one thing we need to consider—when is the earliest the environment will be ready for testing? We need to wait until the CSS and framework of the site are just about done before attempting to show the client anything. Those dates come to a close at the end of May (see Table 8.5), so let's give them as much time as possible before scheduling any reviews. CSS and framework are things that are tested in an on-going manner throughout the project, so I'm not too worried that they won't be ready on time.

The result of these additional tasks can be seen in Table 8.6 under UAT. Notice that the configurator isn't part of the new list—that's because it's the last tool to be finished, so we really don't have any more time on that one. Do you think the client will have any changes after they see these tools? Of course they will! Just remember that code is updated every Thursday, so try to get into the next release with any new code and show the clients their updates.

This is about as far as we're going to take the schedule at this point—but look at the very last line item for cutover (line 130). That little itsy bitsy task seems like it should contain a lot more detail, doesn't it? It

Table 8.5 Looks like the site will be ready to review around May 28

66	Development	82 days	Fri 3/20/15	Mon 7/13/15		
67	CSS	50 days	Fri 3/20/15	Thu 5/28/15	17	FE Dev
68	HTML Framework	60 days	Fri 3/20/15	Thu 6/11/15	17	Web Dev

Table 8.6 Final UAT Schedule

120	UAT	50 days	Fri 5/29/15	Thu 8/6/15		
121	Search	3 days	Fri 5/29/15	Tue 6/2/15	67	Team
122	Video Gallery	3 days	Fri 5/29/15	Tue 6/2/15	67	Team
123	Contact Us	3 days	Thu 6/11/15	Mon 6/15/15	107	Team
124	News Stories	3 days	Thu 6/18/15	Mon 6/22/15	111	Team
125	Trip Planner	3 days	Wed 7/22/15	Fri 7/24/15	115	Team
126	Creative Review	3 days	Wed 7/22/15	Fri 7/24/15	119	Creative
127	Fixes	3 days	Wed 7/29/15	Fri 7/31/15	126	Dev
128	Final Team Review	3 days	Mon 8/3/15	Wed 8/5/15	127	Team
129	Fixes	1 day	Thu 8/6/15	Thu 8/6/15	128	Dev
130	**Cutover**	1 day	Thu 8/6/15	Thu 8/6/15	129	Inf

does and it will. We'll capture all of that activity during the cutover planning and management, but for now we need to wrap up Plan and Define. If the client elected to go with two separate budgets for the project—one for Plan and Define and one for Construct to Close—then we'd be sure to have a formal delivery presentation at this point. Even if we're moving right into Construct to Close, though, we could get a presentation together—it's not a bad idea. Let's think about what that presentation would contain.

WRAPPING UP PLAN AND DEFINE

The goal for our Plan and Define wrap-up presentation is to package up everything we've done so far and present it to the clients so they fully understand and agree with our output, and also so they can, in turn, present the material to their management. Most of the time, the clients like to take a look at the presentation first and then have the agency present it to the larger group, including their management, because we're the experts. If we're taking a sales approach, the goal of the presentation would be to build a case for moving forward with Construct to Close. This isn't generally the role of a PM, though. Our goal will be to simply present the information in an organized and thoughtful manner and to tell the story of how we got here. Along with the presentation, let's hand out binders with all of our pertinent material. Then, our presentation will follow along with the binder content. What should our tabs be? Let's take a look at all of our outputs so far:

1. Gap Analysis Presentation
2. Stakeholder List
3. Business Requirements Document
4. Statement of Work
5. Plan and Define Budget
6. Plan and Define Schedule
7. Sitemap
8. Technical Solution Strategy
9. Wireframes
10. Styleguide
11. Analytics Analysis
12. SEO Analysis
13. Infrastructure Assessment

14. IT BOM
15. Test Strategy & Cases
16. Construct to Close Schedule

Those are all of our outputs, but I'm not sure we should include everything. On the one hand, we want the binder to look robust because the clients probably paid a lot of money for Plan and Define. On the other hand, I could make an argument for not including the following:

- **Budgets:** Be careful with this. Not only do we not want any other agencies looking at our costs, the clients may not want any of their colleagues to see the budget either. If other agencies can see what our pricing strategy is they could use this information to try and steal business from us. And the client's colleagues could complain that our client has too much money to spend (more than they do).
- **Wireframes:** If we include the wireframes it would make the binder *huge*—probably way too big—so I'm tempted to not include them in the binder and just reference them during the presentation.
- **Styleguide:** Again, this would be way too large to put inside the binder. This could be a 200-300 page document all by itself. Let's handle this the same way we're handling the wireframes. Mention them in the presentation but don't put them in the binder. We could always hand the clients USB drives with everything included—even the wireframes and styleguide. In addition, if the client just wanted to pay for Plan and Define, it's possible that we weren't even required to provide a styleguide or even comps at this point.
- **Analytics and SEO:** For the Jetzen project our agency didn't handle analytics or SEO, so it would be weird to present this material as if we created it. I'd ask the client if they want to include this information, and if so perhaps we could invite the people who created the material to join the presentation and possibly cover this for us.
- **Test cases:** We don't have to show the test cases, because the fact that we're producing them is included in the test strategy.

Everything else on our list is good—plus we want to add a little something extra on the end to help transition into Construct to Close. Not to sell it, just to let the team know how we'd approach the rest of the project and what the next steps would be. Here's what I'd cover during our presentation:

Welcome and introductions: This part is easy. Go around the room and have everybody introduce themselves and say which company they work for—make sure to include those who are dialed in on the conference line.

Goals and objectives: The goals for this presentation are to take the participants through a guided history of the project and to communicate options for moving forward. Make sure everybody has a copy of the binder and let them know that we'll be covering each tab, as well as some additional information that wasn't included in the binder due to size. Go through the agenda of topics that will be covered during the presentation. Note that this presentation could last more than one day so don't be surprised if your agenda is pretty lengthy. We've spent months on this and we want to make sure we do a thorough job of explaining the current situation to those in the room who may not have been part of the core team.

Statement of work and schedule: When looking at the 30 rollout steps for this project, the SOW and schedule weren't produced until *after* the business requirements were solidified. But for this presentation, we don't really need to review these materials in-depth and they seem to fit in nicely *before* the BRD, so we're going to cover them up-front. We'll just let the room know that the project deliverables we're about to review were based off of the SOW as authorized by our client, and that the schedule is also included for reference. Again, there's no need to go through these documents, just let them know that they're included in the binder and that this presentation concludes our obligations for the project as described in those documents.

Process summary: Right off the bat, it's a good idea to explain how and why the project was started. Go ahead and tell the story about getting a phone call from your client and how we, right away, put our process in front of them. Show the process on the screen, and explain how the presentation will take us through the outputs captured throughout Plan and Define. As always, open the floor to questions as we move forward, to make sure everybody is following along and understands the material.

Stakeholder list: Reviewing the stakeholder list not only shows the number and variety of people associated with the project, but also provides validity to our outputs by showing that all of the appropriate resources were involved—and contributed according to their expertise. This includes client-side resources as well.

Gap analysis: When determining how much detail to go into for each of these topics, consider who's in the room with us. If the client's boss or senior management is in the room, then we may want to go through the entire gap analysis if they've never seen it before. If everybody in the room has already seen the gap analysis presentation, then just give a synopsis of what we found. This will transition nicely into the business requirements discussion coming up next.

Business requirements and technical solution: This is where our expertise really comes into play, so it's one of the most important parts of the discussion. Here we explain all of the challenges and puzzles we've had to solve in order to provide a solution for the client's desired outcomes. We'll show the requirements alongside their technical solutions to show a complete picture of our analysis, and to help smooth the way toward our next budget for Construct to Close. Remember that for this example, the clients have *not* started development yet, so we need to fully explain what needs to be done and who all needs to be involved from their side. Take the time to ensure that everyone understands what our plans are. I find that when clients decide to pay for a Plan and Define without committing to the whole budget, they know that we're going to find some major challenges in getting this thing off the ground. Usually these challenges involve the technical solution. So, there may be some groans during this part—but suck it up and just put it out there. Don't beat around the bush—be honest about the situation even if it's bad news. We're going to have to deal with it at some point.

Sitemap and wireframes: These two deliverables are normally included in the Plan and Define phase—even if Construct to Close is uncertain. Show the as-is sitemap first, and then show the revised sitemap that our information architects created for this redesign. The sitemaps are easy enough to include in the binder, but as we review the wireframes, explain that they're just too large to be placed in the binder and that a copy has been provided on the supplied USB drive. Of course, the owners of each of these subjects should be presenting their own material, so hopefully the information architect (IA) can take the group through the sitemap and wireframes on-screen.

Styleguide: I don't want to hurt anybody's feelings, but nobody will want to go into a deep study of the styleguide. It is important, though, to explain the *concept* of the styleguide. Styleguides are prepared not only for the initial build of the website, but also to serve as a guidepost moving forward. Any new functionality or creative should be held up to the styleguide for compliance. Plus, it cost a *ton* of money so we should at least mention how important it is to the process.

Analytics and SEO: We had to include both analytics and SEO in Plan and Define in order to come to a final Construct to Close budget number. Any delta in the analytics data that will be captured with the new website design should be mentioned, as well as any SEO changes. Also crucial to mention are the expected increases in the client's key performance indicators—probably a major reason they're engaging in this process in the first place.

Infrastructure assessment and bill of materials: Hopefully this part is short, because if it isn't, that means there's most likely a problem that may be expensive to resolve. We want to be able to say that the current infrastructure and bill of materials can handle any extra weight expected from the new website. If not, we have to explain what needs to be purchased and how much that might cost. Of course, with the Jetzen project, this information will come from their own people, since they're hosting the site internally—but I'll bet they asked for some guidance from the agency during their assessments.

Test strategy: This is an important sell tactic for Construct to Close. I don't think these guys would have hired us in the first place, if they didn't believe we had world-class test strategies in place, but reiterating the nonfunctional testing policies and procedures one more time will help build the client's confidence that we've done this before and we're ready to proceed. It also provides a nice transition to our next subjects—which are targeted at securing the needed approvals to move forward with Construct to Close.

Construct to Close schedule and budget: Not much to explain here. Let's put on our sellin' shoes and make it easy for the clients to say the word *go*. Please note that we may need to prepare a whole new SOW for the second half of this project since there will be a new budget. Remember—be careful about this and strategize with your client on the best approach for moving forward with this part.

Next steps: Clearly, the next steps for the client are to sign their approval and give us the go-ahead to move forward. Or, maybe during the meeting some questions came up that we need to follow up on, before they can provide an answer. If they're all ready to go, then the next steps for us will be to get ready to receive assets from creative and to prepare the team for a whirlwind. Let's do this!

In this chapter we:

- ✔ Got the test strategy approved
- ✔ Had the test cases written and approved
- ✔ Received the package delivery dates from the creative team
- ✔ Finished the Construct to Close schedule
- ✔ Wrapped up Plan and Define with an awesome presentation

SECTION III:
CONSTRUCT TO CLOSE

9

CONTENT

It's time once again to take a look at our process map and see where we're at. Table 9.1 shows that we have a lot more check marks in there now. It feels good to have the planning part over so we can concentrate on executing, doesn't it? Why does looking at those check marks give me such a good feeling of accomplishment? Well, let's make some more! In this chapter we're going to get ready to take delivery of the assets, check to make sure the assets match specifications, enter them into the website, and then do a page-by-page quality review.

After the content tracker is created, Steps 22–24 of the rollout process will happen sequentially as the assets are delivered according to the packages previously identified. The asset manager will take delivery of the assets, check them, and then send them on to either content entry or back to the appropriate party, if there are errors that need to be corrected; all the while constantly updating and distributing the content tracker to the team to keep everybody informed of the latest activity.

CREATE CONTENT TRACKER

When the creative team starts their production work, they'll probably ask for a content tracker to help them make sure that they deliver all the necessary assets for each page of the website. This document will also help our internal asset manager keep track of everything we're supposed to receive, what has already come in, what stage everything is at in the process, and what's still outstanding. The tracker is a spreadsheet, as shown in Table 9.2, which documents the asset type, description, width,

Table 9.1   Check marks on our rollout process

PLAN AND DEFINE	
INITIATING	**PLANNING**
✓ Gap Analysis	✓ Plan and Define Schedule
✓ Workshop	✓ Communication Planning
✓ Stakeholder List	✓ Sitemap
✓ Business Requirements Document	✓ Technical Solution Strategy
✓ Preliminary Budget Estimate	✓ Wireframes and Functional Specifications
✓ Statement of Work **(Tollgate)**	✓ Styleguide
	✓ Analytics Analysis
	✓ SEO Analysis
	✓ Infrastructure Assessment
	✓ IT BOM
	✓ Development and Change Management
	✓ Test Strategy and Cases
	✓ Package Identification
	✓ Construct to Close Schedule

CONSTRUCT TO CLOSE
21. Create Content Tracker
22. Asset Quality Review
23. Content Entry
24. Quality Assurance
25. System Integration Testing (SIT)
26. User Acceptance Testing (UAT)
27. Nonfunctional Testing (Performance, Security, Disaster Recovery, Failover)
28. 301 Redirects
29. Cutover Management
30. Transition to Operations

Table 9.2 The content tracker

Home				
Asset Type	**Description**	**Width**	**Height**	**Status / Comments**
Text	Copy	N/A	N/A	
png	Jetzen Logo	95	45	
png	Microscope	15	15	
png	Masthead Arrows	20	20	
jpg	Learn More Button	110	25	Missing hover state
jpg	Masthead Images	960	N/A	
jpg	News Story Teasers	230	N/A	
jpg	Special Events	470	N/A	
png	Social Icon 1	34	34	Missing link
png	Social Icon 2	34	34	Missing link
png	Social Icon 3	34	34	Missing link
Copy or Comp	Sitemap	N/A	N/A	

height, and status. Sometimes it will also list the maximum weight for each image, which is based on image size. Most of this information can be found in the styleguide, and was based on an elaborate content strategy that the team designed for our responsive website.

The information on the tracker example in Table 9.2 follows along with the list of homepage assets we came up with in the last chapter. Remember when we listed all the assets needed for the homepage based on our wireframe? Well, we put them all on the tracker and now this document can be used as a status report for the assets. We'll send it out as often as needed in order to keep everybody updated on our progress. Asset delivery also impacts our development timeline so we need to make sure that they're arriving on time and according to specification. At this time I'd like to go through each of the columns in the spreadsheet and provide a detailed description of each, including where one would find the information.

Asset type: Types of assets can be an image (.jpg or .png), video, text, pdf, excel or any other file type we expect to receive from the creative team. I'm actually starting off with a hard one here because the asset type for images isn't always documented in the styleguide—sometimes we have to guess at it first, and then run it by the creative team for confirmation. It doesn't *always* matter if the team delivers a .jpg or .png, but there are times where it will make a big difference. It's fine to put .jpg/.png in there when it doesn't matter—and then for those assets where we definitely need a .png we'll specify it in this column. The asset manager can usually take a look at the comps and figure out what type of image we'll need, but sometimes we just go ahead and ask the creative team to update the tracker with the types of assets that we should expect so we don't have to guess.

Description: This clarifying column should contain a concise label that would allow anybody to read and understand what the asset will be used for. Notice from the example in Table 9.2 that sometimes we list out reoccurring items separately and sometimes we bunch them together. For instance, for the masthead images, we weren't too sure how many the creative agency was going to hand over, so we just bunched them all together in one row. It doesn't really matter how many of those we get so there's no need to call them out separately. The masthead can swipe through as many or as few images as they want—or, the creative team can put just one up there if they elect to take that route. On the other hand, for the social icons we're definitely expecting three of those—the homepage wireframe and comp have always listed out three particular icons and we want to make sure we are looking for, and then receive, all three.

Dimensions: Asset dimensions can be found in the styleguide and are valid for images only because videos will normally fill to the size of their container. Notice from the content tracker example that sometimes we list both the width and height—and sometimes we don't. That's because the height doesn't matter as much as the width does. This takes us back to the grid discussion from the information architecture section of this book (see Figure 9.1). Widths are very important to be accurate about because we only have so much space to work with and the components in which the images will live are probably coded to only go so far left or right. We measured out all the different scenarios (two columns versus four columns) when we created the grid system during the initial development of the styleguide to make sure we were all aligned on the plan. All comps should have been developed based on this system, so there shouldn't be a problem, but sometimes the measurements are a little off so they must be checked upon receipt.

Conversely, heights are not as much of a problem because the user can scroll down as much as they need—there are just a couple of things to be aware of regarding image height—live area and design aesthetics. The following examples I've provided are based on desktop and tablet but can also be applied to mobile.

Live area: This is the portion of the website that the user can see on-screen without having to scroll. If the masthead is too long, then it could

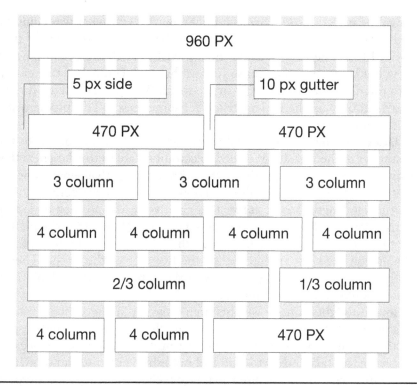

Figure 9.1 The grid system was put in place to provide accurate measurements and guidelines for both the creative and development teams

take up the whole screen without showing anything underneath. Maybe this doesn't seem like a big problem, but I ran into this just recently and it was a really strange user experience. The site I was on was basically hundreds of pages of frequently asked questions and answers about a certain book. The user has the option in the masthead area to click various links like *Top 20 Questions* or *Question of the Week*, but most users have specific questions in mind that they're interested in and they type their questions into the search bar. I had been to the site before and had used it for years, but they just updated their design and now the masthead area took up the entire live area. So, when I first used the site with the new design, I typed in my question and pressed *search*, but nothing happened. It was so confusing—previously, the page would change, and a list of search results would appear. I thought the site wasn't working right, until I figured out that it was all happening below the live area and I just couldn't see it. When I finally scrolled down (actually swiped down because I was looking at it on a tablet), I saw what was going on and decided to contact the organization because I wasn't sure if they noticed this. The page worked fine when I held the tablet in portrait view, but landscape view was so strange that I felt compelled to mention it to them, because I love their site so much. They actually fixed it right away, but only on the tablet version for some reason. This problem still exists for desktop.

Design aesthetics: Figure 9.2 shows why length is important to the aesthetics of the website. We need to communicate to the designers that length is not a fixed specification without saying that length doesn't matter. Most of them will already know this, but I have definitely said that image length *doesn't matter* to a creative team and then ended up with something like we're seeing in Figure 9.2. Notice that the picture on the left looks too short, the one in the middle looks about right, but the one on the right extends below the live area of the screen. There's nothing technically wrong with any of them, but they don't look right. They should probably all line up.

Image weights: Image weights aren't shown in the content tracker example, but it's something important to check for and should be part of the overall content strategy. The reason it's so important is because heavy images take longer to load and publish than lighter images. Table 9.3 shows what might be used as image weight guidelines for a typical responsive website. Based on the size of the image, there's a maximum allowable weight that the designers have to stay within, but don't worry because they have several techniques they can use to optimize the images in order to reach these goals. The discussions surrounding image optimization have become even more important since the advent of responsive websites—because we need images that look great on a big screen but are light enough to load onto a mobile phone. Part of the content strategy might even be to switch to a different image altogether at a certain viewpoint or percent of maximum width.

Status: With all of the different assets coming in and the rigid quality review process that they need to adhere to, one can see why the content tracker is a vital part of the process. Some people think it's just double the amount of work, because all of the information is already contained in the styleguide, but the styleguide doesn't provide page-by-page directions or allow for monitoring. The status column for the tracker can be used to show if the asset was delivered on time, whether or not it passed

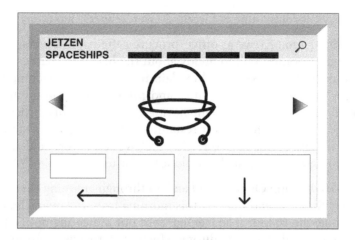

Figure 9.2 The *live* area is the area on the screen that users can see without scrolling

Table 9.3 Image weight restrictions

Image (in pixels)	Max. weight
50 x 50	10k
150 x 150	15k
300 x 300	30k
500 x 500	50k
Any larger image	100k

asset quality review (AQR), what exactly about the asset made it not pass AQR, or what stage the asset is in. For instance, are they still working on it in content entry or is it already loaded and in testing? Think of this as a way to communicate to the creative team about what they should be working on. We need to create a tracker for every single page and every single element of the website. Again, the asset manager puts this together and updates it as the assets are delivered.

ASSET QUALITY REVIEW

By now the theory of what an AQR is should be pretty clear, based on what was described for the content tracker. So let's go through a day in the life of an asset manager:

Monday Morning, June 8: Group 2 asset delivery due today

9:00–9:30 a.m.: Arrive at office, procure caffeinated beverage, and turn on computer. Talk to buddy about the big game that was televised over the weekend and explain how poor the coaching was. Listen to his weekend activities and pretend to be interested in how he spent Saturday night having dinner with his girlfriend's college friends at the newest restaurant downtown. Brain starts to wake up and suddenly we realize that we're at work and should start to figure out what the to-do list is for today.

9:30–10:00 a.m.: Open e-mail and go through anything that came in over the weekend or early this morning from the overachievers. Roxanne, the rollout manager, has sent out a meeting notice for a conference call at 2 p.m., during which we'll review the status for the Group 2 assets that are due today. We're not expecting anything to actually arrive before 5 p.m., but we want to make sure everything will be delivered at least by the time we arrive at the office tomorrow morning.

10:00–10:45 a.m.: Update the content tracker with the latest activity that happened over the weekend or this morning. We received the destination URLs for the three social icons located in the global footer. Drop them into a browser to make sure they are going to the right spot—and they are—so we mark down on the tracker that they passed AQR and were sent along to content entry. Also, the search engine optimization (SEO) team updated the homepage copy doc with the alt tags that were missing for News Stories. Everything from Group 1 seems to have been delivered and has passed AQR just in time for Group 2 to be delivered. Send out the updated tracker and make this announcement, congratulating the team on a job well done.

10:45–11:00 a.m.: The creative lead calls to let us know that they don't have access to the videos that are supposed to be delivered with Group 2. They have no idea where the videos are, but they've definitely seen them before. That's not helpful. Anyway, this is told to Roxanne, who will call the clients and ask where the videos are housed. The good news is that the creative lead indicated that he was about to upload a partial delivery for Group 2 and that he'd send an e-mail when it's ready.

11:00 a.m.–12:00 p.m.: Wondering how soon those assets will actually be uploaded. Decide that it'll probably take a while, so start to go over the content tracker for Group 2 and then organize the online file folders to get ready for delivery. A phone call comes in from one of the other project managers—they have a question about another project that they need our help with. There's a big incentive launch this weekend and the assets are expected to arrive on Friday, which doesn't provide much time to get things built, reviewed, and launched. We mark down the details and ask for any comps or wireframes that are available so we can start to get ready for that project while waiting for Jetzen's Group 2 assets.

12:00–1:00 p.m.: Lunch. It's nice out, so a bunch of people are walking over to get some pizza.

1:00–1:15 p.m.: Roxanne calls back and says that the videos are on YouTube and is hoping we can just stream them in from there, because nobody knows what happened to the originals. She's going to get back to me.

1:15–2:00 p.m.: Get an e-mail from the creative lead saying that the partial delivery of assets should be ready. Go to the delivery location, but don't see anything. Call the creative guy back about discrepancy and figure out that he put them in the wrong spot. While he's moving the assets over, we start to log what's coming in so we can report out during the 2 p.m. call.

2:00–3:00 p.m.: Time for the conference call. Go through everything we saw being delivered and tell everybody that the next day or so will be spent going through the delivery and updating the tracker. The rollout manager confirms that we'll be able to stream in the videos as long as we can get the YouTube account information from the clients. Clients are silent at first, but then promise to get the information as soon as possible. The creative team says that they're just making a few final adjustments on the rest of Group 2, and will have them uploaded by the end of day. Then Roxanne goes through a few items on her status report and confirms the next steps for Jetzen.

3:00–3:15 p.m.: Get an afternoon snack and something to drink. Begin to wonder when caffeine became a necessity for getting through the day. Think about vacation coming up in July.

3:15–11:59 p.m.: Download assets and conduct AQR against content tracker. Log all results, and send approved assets along to content entry as they're ready. During the evening, check email periodically to make sure that the assets get delivered, and they finally are at 11:25 p.m. That's fine—as long as they were delivered before tomorrow morning. Send a note confirming receipt and thanking the team for making this important deadline. Sleep.

The asset manager has to be very meticulous about assessing every single asset, including SEO and copy. They have to size the images, weigh them, check the asset type, and review them against the comps. Then they have to open up the copy docs to make sure they understand how the copy follows along with the comp and assets, and also to make sure that all the SEO is there. If there's a problem, the asset is sent back to the appropriate party for fixes, and marked down as *delivered but did not pass AQR* due to whatever the issue was. There's so much going on and so much to keep track of—things can really get out of control quickly without this key resource in place. But, once the assets pass AQR, they should be fairly easy to input because we've spent so much time upfront planning things out.

Content Entry

There doesn't seem to be a lot to write about on content entry, does there? Enter the content—done. There's some truth to that, but I do want to go over a couple of things, just to provide a complete picture. First let's talk about digital asset management (DAM). The DAM is a central repository where all of the assets for our web project are stored and from where they are pulled into our website for publishing. Whether we're using a content management system or if we're creating a traditional html site, images are referred to, or linked to, from the main page and pulled in. Code for pulling in images might look something like this:

```
src="http://www.jetzendam.com/content/home/2015_
Atlantic_Star/Features/beautyshot-atlantic-star-960x320
.jpg"
```

What if we needed to update that image we just referenced? If we don't change the name of the asset from "beautyshot-atlantic-star-960x320

.jpg" then there's a good chance that, as soon as it's added to the DAM, it will be pulled into the site because the system is going to just replace the old image just like it would if you were saving an image to your desktop. We might not get a cautionary note that says something like: *An image with this name already exists, do you want to replace it?* It might just go live right away, depending on the asset manager system that's in place—and if that's the goal, then that's great—keep the name. But if we're saving this picture for a special launch initiative, then be careful to rename it. Making a simple change like removing the extra hyphen between *atlantic* and *star* (beautyshot-atlanticstar-960x320.jpg) will differentiate it enough so it won't get pulled into the site, while still allowing us to keep the original SEO.

Figure 9.3 shows how we might structure the DAM for Jetzen. It doesn't technically matter that much how we structure the folders, but having an organized structure does make a big difference in our team's ability to maintain and manage the site. Each of the line items seen in Figure 9.3 represents a folder that houses various images and files re-quired for individual pages. One major reason it's so important to keep the structure organized is in case of an incident. What if an image is no longer showing on the screen and we have to find it, to see why it's not working? The task of finding the right image is extremely difficult without a folder structure in place that matches the organization of our

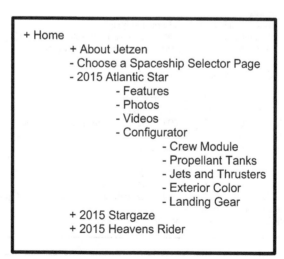

Figure 9.3 The folder structure for the DAM should be well organized

sitemap. Look at the file reference provided previously and try to figure out where that image would be located within this structure. See—I keep coming up with fun little things to do, right?

Another important thing about the DAM is to keep it clean. I know that nobody likes to delete images for fear that they're being used somewhere on the site and we'll get in trouble for breaking something. So yes, we need to be extremely careful, but there's also a very strong case to be made for going through the DAM every year or so and cleaning it up. Delete the images and files that are no longer in use because all of this added weight on the system can slow down our website. And, it's just messy which makes it difficult to find what we're looking for.

Next, let's take a look inside one of the copy docs that came in with Jetzen's Group 1 delivery—specifically for the homepage. Figure 9.4 shows that the SEO team has started off by providing the homepage URL (as if we didn't know that one) and the meta description. Remember that this description will be what users see on their search engines

---

HOMEPAGE V1
June 1, 2015

[URL: www.jetzenspaceships.com]
[Meta Description: Jetzen Spaceships manufactures and sells a full range of space travel and exploration vehicles.]

Masthead
[MH Image 1: JS_Atlantic_Hompage.jpg]
[NAMING CONVENTION: 2015-jetz en-atlantic-star-960x320.jpg]
[ALT TEXT: 2015 Atlantic Star from Jetzen Spaceships]
[CTA Button] Learn More
[LINKS TO: www.jetzen.com/atlantic-star]

[MH Image 2: JS_Stargaze_Homepage.jpg]
[ALT TEXT: 2015 Stargaze from Jetzen Spaceships]
[NAMING CONVENTION: 2015-jetzen-stargaze-960x320.jpg]
[CTA Button] Learn More
[LINKS TO: www.jetzen.com/stargaze]

[MH Image 3: JS_Rider_Homepage.jpg]
[ALT TEXT: Heaven's Rider from Jetzen Spaceships]
[NAMING CONVENTION: 2015-jetzen-heavens-rider-960x320.jpg]
[CTA Button] Learn More
[LINKS TO: www.jetzen.com/heavens-rider]

Figure 9.4   This copy doc shows that each of the three available spaceships will be featured in the homepage masthead

as the short descriptive copy about the website. Then the copy doc goes right into the masthead instructions. By reading this we can tell that the creative team has provided us with three images that they want rotating in the masthead area, along with what their file names are. *JS_Atlanti _Homepage.jpg* is the name of the first masthead asset that we'll be looking for in the delivery folder, and the SEO team wants us to rename it to "2015-jetzen-atlantic-star-960x320.png".

Next, we see that the alt text has been provided to us along with the call-to-action instructions. The creative team is asking for a button on the image that says *Learn More* and that button should link to the Atlantic Star features page. If they want us to link to the Atlantic Star photos page, the URL would say something like *www.jetzen .com/atlantic-star/photos*. Copy inside of brackets indicates that they're giving us instructions for the content team and not copy that needs to be put on the website. For this document, the only copy not placed inside brackets is the copy for the button. The developers or content authors can make the button say pretty much whatever the creative team wants, so the copy doc has to indicate what words to use. The asset manager would probably check this against the comp to make sure that both things match—and if they didn't, they'd ask which one was correct. We're always instructed to go by the copy doc, but common courtesy (and the fact that we don't want to do it twice) demands that we ask—just to make sure.

Figure 9.5 shows an example of a main headline for the page and some supporting copy. Looks like under the main headline copy there's going to be a button that says *Find a Dealer* that links to the actual application that the developers are working on, and then underneath that there's a sub-headline. Within the body copy for the sub-headline are instructions for where to hot-link the underlined words. The content team has to be cautious not only to make that connection, but also not to include the instructive copy in the website.

Finally, we have Figure 9.6 showing what some disclaimer copy would look like. The content team should copy and paste the text into the site exactly as shown. According to the comp, the disclaimers go just above the global footer. The global footer information just has to be supplied one time as it's the same on every single page, and it wouldn't be uncommon for the team to include it with the homepage delivery. But—there's a problem. Notice that the social icons are missing the destination URLs. Remember from the beginning of this chapter that the asset manager caught this and included the error in the tracker?

> **H1 Headline: GET READY FOR SPRING FLIGHTS!**
> [BODY]
> This is just dummy copy written to represent something one might find in a copy document. Remember that H1 refers to the most important headline on the page.
>
> [CTA button] FIND A DEALER
> [LINKS TO: www.jetzenspaceships.com/find-a-dealer]
>
> **H2 Sub Headline: PICK UP A SPARE BATTERY NOW!**
> [BODY]
> H2's are sub-headlines and will usually be a smaller font size than the H1 headline. The paragraph text might contain text that is <u>hot linked</u> to another page [LINKS TO: www.jetzenspaceships.com/owner-resources/service#OwnersManuals] so we must provide the destination URL in the copy.

Figure 9.5   H1 and H2 headlines use different font treatments which were described in the styleguide and are now coded into the Cascading Style Sheet

> **LEGAL DISCLAIMERS**
> * This is just dummy copy written to represent something one might find in disclaimer copy.
>
> 1.   This is a fake disclaimer.
>
> Effective 6/1/15: Service replacement parts, including batteries, are covered for the remainder of the spaceship's warranty, or 12 months with unlimited miles, whichever occurs first.
>
> **GLOBAL FOOTER**
> [Facebook Icon: facebook.png]
> [NAMING CONVENTION: jetzen-facebook-34x34.png]
> [LINKS TO: TBD]
>
> [Twitter Icon: facebook.png]
> [NAMING CONVENTION: jetzen-twitter-34x34.png]
> [LINKS TO: TBD]
>
> [Instagram Icon: facebook.png]
> [NAMING CONVENTION: jetzen-instagram-34x34.png]
> [LINKS TO: TBD]
>
> **[POP UP BOX] Copyright and Trademark Information:** Pretend there is copy here that relates to copyright and trademark information.
>
> **[POP UP BOX] Privacy Statement:** Pretend there is copy here that relates to privacy statement.

Figure 9.6   All text in the document should be copied and pasted into the website to avoid error

So the asset manager has to read through the copy docs, make sure they understand all of the instructions, check to see if they align with what they're seeing on the comps, and verify that all of the assets referred to in the doc are accounted for in the delivery folder. A good asset manager makes sure that they're able to answer any questions that the content team might have before sending along the assets to them. If they're unsure about anything, they simply ask the creative team for clarification. Once the assets pass AQR and are entered into the website, they're published so that the test team can make sure everything was done correctly.

QUALITY ASSURANCE

At this point in the process, our quality team is performing content testing versus systems integration or user acceptance testing—which will be discussed in detail in the next chapter. Once the content team finishes each page, they'll be passed along to the test team who'll check the work against the comp and copy doc. The test team should be able to find all of those places where the instructional copy in brackets was accidentally left in, or if, for instance, the buttons are not clicking through to the correct destination. We're going to hit testing pretty hard in the next chapter, so I don't want to go into too much detail here. My last remark is to make sure the quality team knows that it's not their job to find proof errors in the copy. Their job is to make sure that the copy was taken directly from the copy doc without alteration and placed in the website. Once they start pointing out proofreading errors *just to be nice guys* then all of the sudden they're on the hook for it if something slips by. Don't set this expectation with either the clients or the creative team. All copy should have been proofread and approved by legal, before being delivered to the production team. Maybe if the creative agency and the production agency are one in the same we can ask them to point these things out, but again, it's not their job. They're not editors, they are quality professionals.

Now that we're starting to see published pages, the project is getting really exciting. I find it amazing to be part of a digital project from concept to delivery, because what used to be thoughts or ideas in somebody's head are now real things—things we can look at, read, listen to, swipe at, and otherwise interact with. It's really the aspect that makes advertising and marketing such a fun career choice—watching ideas come to life—not to mention making something that improves people's lives. It may be difficult for some people to come to terms with how their jobs in marketing or advertising can improve lives, but it was taught to me at a young age, so I totally buy into it. I mean, we're not delivering hearts on ice here, but we should still do our jobs to the best of our ability and for the purpose of helping others. Here's a few ways we are helping people through the Jetzen project:

Making our colleagues lives easier: The way I see it, if I do my job well, then that's helping other people because my job as a project manager really affects other people's stress levels. I try to be kind, nice, and helpful—even when other people aren't.

Putting food on the table: Think about how many people around the world would be involved with a company like Jetzen. Well, we said back in Chapter 1 that Jetzen has 19,273 employees. If we do our jobs well, the result should be an increase in spaceship sales. That keeps people employed and prospering. The more money they make, the better they're able to provide for their families and, in turn, help others. In addition to the employees of Jetzen, there are large numbers of parts suppliers, fuel suppliers, and other support teams that are affected by how well we do our jobs.

SOMETHING EXTRA—FAVICONS

Don't forget the favicon! This is something pretty easy to forget because it isn't crucial for the design, development, or deployment of the site. Favicon stands for *favorite icon* and they're the little mini logos that can be found on the browser tab of many websites (not everybody uses them but, they're still a nice thing to have). Also, they often pop up in bookmarks and sometimes in address lines, depending on which browser is being used. Ask the developer what type and size of icon they need, make the request to the creative agency, and then send it along.

In this chapter we:

- ✔ Created the content tracker
- ✔ Started our asset quality review
- ✔ Started content entry
- ✔ Began the quality analysis of our content pages

10

TESTING

Now that our developers are finishing up their projects and we've done a bit of unit testing, we're ready to start executing against the test strategy that we put together. First we'll complete systems integration testing (SIT), and after that we'll move on to user acceptance testing (UAT) and nonfunctional testing—just like we explained in the test strategy document. Katie is the name of the lady who'll be responsible for testing on this project, so we need to make sure she knows that our project is running on schedule and that we'll be ready to start SIT around March 27—for search, as expected.

SYSTEMS INTEGRATED TESTING

One of the first things we'll work with Katie on will be to customize the ticketing system and create a reporting dashboard, so we can easily sort and review our tickets. I have no doubt that the reader understands what a ticketing system is, but just in case, I'll provide a quick definition. Every digital agency has a ticketing system to manage defects and change requests. Smaller agencies can use open source ticketing systems that can be found on the Internet, and larger companies buy into more robust systems which can be customized for their particular needs. Figure 10.1 shows an example of what a ticket in one of these systems might look like, so let's review the fields in this figure.

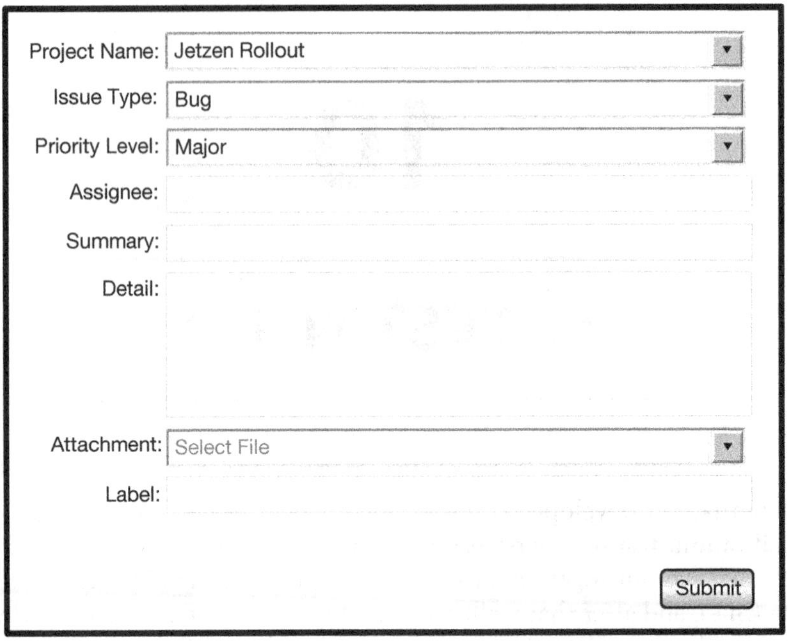

Project Name:	Jetzen Rollout ▼
Issue Type:	Bug ▼
Priority Level:	Major ▼
Assignee:	
Summary:	
Detail:	
Attachment:	Select File ▼
Label:	
	Submit

Figure 10.1   Most ticketing systems, or bug trackers, use the fields shown here to help clearly define and manage issues

Project: For this project we've added *Jetzen Rollout* as our project title.

Issue type: It's important to accurately select the issue type because *changes* may be less critical to fix than *bugs*. A *bug* is when the product does not match the specifications or does not meet the testing standards that were previously established. A *change* indicates that whatever we're asking to be fixed was never specified in the first place. Remember back when we were creating the business requirements document (BRD) and I made sure to mention to the team that they should be very specific in their requirements? I said, "If it isn't listed in the BRD, don't expect it to be delivered." This is why I made sure to emphasize that. In one person's mind, it may seem like standard procedure to make sure the customer's e-mail entry in the contact form is checked and verified to make sure it's valid, but if nobody told us to do that, then we wouldn't have. Maybe the client didn't tell us to verify the e-mail because they didn't want us to. We need to make sure we closely review what's being sent through as *bugs* and make sure they're not *changes*, because most changes should be put off until after launch (Phase 2) if possible. Our mission right now is to deliver the original specifications.

Priority: This field is automatically populated with *major* status which is accurate for most bugs. Also available are *critical* and *minor*. Critical is used to indicate that the issue is either blocking us from further testing or that the site can't be launched until this issue is fixed. We have to start thinking in those terms because there may very well be issues still not fixed before launch, and the client will have to decide what the deal breakers are. Minor issues can wait until after launch, but they're so easy to fix that we can probably knock them out of the way pretty quickly.

Assignee: When opening a ticket, it should be assigned to the specific resource we think can fix the problem, or to a resource that needs to provide assets or other information before the problem can be fixed. So many people forget to assign the ticket to somebody, so they end up just sitting out there—doing nothing. When there's nothing else to do, I like to just click around to make sure tickets are moving.

Summary: Provide a brief summary of the issue. Because there are so many different projects all happening at the same time, we may want to start every summary with "Jetzen Rollout" followed by some specifics. For example, *Jetzen Rollout—Contact form missing copy on mobile* would let us know that there's some copy missing in the mobile version of the contact form, although we don't know exactly where or what until we read the description.

Description: Explain the issue with as much detail as possible so that anyone can read the ticket and understand where the issue exists and what exactly is wrong. It's *always* necessary to provide a URL address for where to find the issue, and *usually* necessary to provide the steps to recreate the issue. For instance, if one of the error messages on the contact form is missing some text, then we should provide the steps to take in order to make that same error message appear so that the person who is responsible for fixing it can easily locate the issue. For example, after following the steps listed below, the user should get an error message because the comment section is a mandatory field on the contact form.

Step 1:   Go to *www.jetzen.com/contact* on a mobile device

Step 2:   Complete all open fields, but do not input a comment

Step 3:   Press submit button

Attachment: All tickets should have a screen shot attachment which clearly shows where and what the issue is. In the case of the missing copy on the contact form, the tester should take a screen shot of the incorrect error message and circle or highlight it in a way that would be clear to the person fixing the issue. Also, it would be nice if they would let us know which copy is missing. Perhaps this is located in the wireframes or the data mapping spreadsheet. I hate it when testers mark down that copy is missing, but they don't write down what the missing copy is—then we have to go look for it. I mean, they *must* know what the missing copy is, or else how did they know it was missing?

Labels: It's super helpful to use labels on tickets so we can then sort them based on those labels. Since we want to have a dashboard for this project, we'll ask everybody who opens a ticket to enter the label *Jet15* in the label field. This will bring up every single ticket associated with this project, as long as we put that label in every single ticket. Sometimes I like to add a label of my own, so that later on I can pull up all the tickets that I opened up for a particular tool or subject. Labels are available just to help sort and manage all the tickets. I know we don't like to think that there will be a bunch of tickets, but there will be—that's just how it is.

The first couple of tools that will be ready for SIT are search and the video player. Search shouldn't be much of a problem, since our agency has built tons of search tools for other websites and they all pretty much work the same way. We just need to make sure it's picking up Jetzen's Cascading Style Sheet (CSS) and that the results are accurate. The test team will conduct a few searches and document the results. They'll type in some things that the search tool should find within the site, such as *Stargaze* or *Spaceship*, and then a couple of things that wouldn't be found within the site, to see how those results are posted. If everything seems to be working fine then we can give it a *pass*. According to the schedule, here are our testing start dates for each tool:

Search:	March 27
Video Player:	May 8
Contact Form:	June 5
News Stories:	June 12
Trip Planner:	July 16
Configurator:	July 17

June looks like a pretty busy month, with both the Contact Form and News Stories being ready for testing. Figure 10.2 shows the dates from our project plan put into a calendar format. I know this is a pain, but I do sometimes drop project plan dates into a calendar even though it's

Week of 6/1/2015				
6/1/2015	6/2/2015	6/3/2015	6/4/2015	6/5/2015
			Contact form released to pre-production	Contact form internal testing
Week of 6/8/2015				
6/8/2015	6/9/2015	6/10/2015	6/11/2015	6/12/2015
Contact form internal testing	Contact form code fixes	Contact form code fixes and package prep	News stories released to pre-production and contact form updated.	News stories internal testing Contact form E2E testing
Week of 6/15/2015				
6/15/2015	6/16/2015	6/17/2015	6/18/2015	6/19/2015
News stories internal testing	Code fixes to both	Code fixes to both and package prep	Code updates in pre-production	Re-testing on both continues

Figure 10.2 We'll be busy in June with at least two tools going through SIT

redundant because calendars are easier to read and comprehend. Since we have to work with people from Jetzen's IT department to get the tools and forms tested from end-to-end, I want to make things as clear as possible.

As shown in the calendar example in Figure 10.2, we'll have a lot going on in June. And this doesn't even take into account the unit tests that are going on for the Trip Planner and Configurator, follow-up tests for Search and the Video Player, and all of the content coming and going. The calendar shows that every Thursday, as planned, the preproduction environment will be updated with the latest code for our testers. Then the test team will have all of Friday and Monday to conduct their tests, using the test cases they've prepared in advance—and on Tuesday we should start making sure the developers have enough time to make the fixes and update the package for next Thursday's release. Notice that I have the contact form going through (at least) one round of internal tests before we show it to our client's IT department for the end-to-end (E2E) testing. We want to try and catch anything in advance of the client's looking at it—even though we still have plenty of time before UAT.

End-to-End Testing

For forms that require E2E testing, there's a trick to catching potential flaws before we expose our code to the client. The developers at our agency actually develop the xml, so let's have somebody fill out a form and then pull the xml to see what data is being captured. Does it match the specifications put into the data mapping spreadsheet we put together in Chapter 7? Make sure that it does, because the database on the other end is expecting it to, and if there's any deviation the data won't make it through. It's important to check this in advance, because we don't want to be on the phone with Jetzen IT while we're sending through a form and then have something on our end be wrong that should have been checked ahead of time. There are a couple ways to conduct E2E testing, but Figure 10.3 shows the basic flow of data from one end to the other. There's likely to be several different database touch points before the data gets to the final destination—depending on where the data needs to end up.

1. **Live on the phone:** Back when we discussed our friend *bit* in Chapter 5, we mentioned that he could end up in one of several different locations, depending on what the customer is looking for—thus, we need to test every scenario. For instance, if we were testing to make sure bit made it through to the West Zone Office, we'd start up a conference call with the person who will be filling out the form in our test environment, the developer who can check the xml capture, the person or *persons* who can check each database to make sure bit is arriving and leaving those locations, and then, the person in the West Zone Office

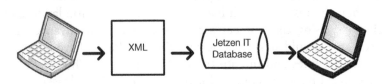

Figure 10.3   Follow *bit* as he moves through the system

who would be responsible for receiving the customer contact and responding back as necessary. If anything goes wrong along the way, we have everybody on the phone to help sort through what's going wrong.

2. **Through e-mail:** We can conduct these same tests through e-mail, but we'll lose the ability to troubleshoot in real-time. If everything goes smoothly, then there's no harm done, but if bit doesn't make it through, we'll spend a lot of time sending e-mails back and forth and then waiting for people to respond. And remember—we don't have a lot of time to get these tests done.

So we go through this process for each of the tools, while maintaining traction with the content team and all of the various problems and roadblocks that come our way. At this point part of the rollout manager's job will be to make sure our resources have our project at the top of their minds, because they're probably working on several other things at the same time. Don't be shy about pouring on the charm to keep your resources engaged in the project.

Troubleshooting

Regarding that missing copy from the mobile form, where should we start looking to figure out what's going wrong? Well, if I am assigned that ticket, the first thing I'm going to do is to verify the error. First, I'll go to the desktop site and see if the copy is missing on that version. I enter *www.jetzen.com/contact*. Well this is strange—the copy is fine on the desktop page—maybe the testing person made a mistake on this one, because the copy for both desktop and mobile comes from the same place. Let's check the mobile version. To look at how the same page comes across in mobile, I have to drag the right edge of the screen toward the left until I see that it goes down to one column. So, I check that out and, yep...the copy is indeed missing. Dang it, I wanted to put *could not reproduce error* in the ticket and close it out! This is strange. Maybe there's something in the code. The content strategy has the copy dropping down and wrapping as it moves from a wide screen to a smaller viewport, so maybe there's something happening with that. I'll send the ticket along to the developers and have them take a look.

This is one of the reasons I love digital—it often comes down to solving puzzles and using our brains to figure out problems. I really enjoy using all of the web developer tools available as browser add-ons and working to figure out what's going wrong. Some people like that stuff and others don't. At any rate, we need to try and get all of the known issues solved before UAT begins.

USER ACCEPTANCE TESTING

Wow, we're getting dangerously close to cutover, right? SIT is finishing up and we need to prepare for UAT with the external team. The first thing we should do is to meet internally and come up with a plan of action. There are a lot of different things to consider, and a lot of different ways we could do this, so let's make sure we're all on the same page. Figure 10.4 shows the agenda we'll use for our internal UAT prep meeting.

UAT schedule: The goal of this meeting is to plan out how we're going to conduct UAT so that the process is clear for everybody involved and controlled from the agency side. People can get really hyped up at this point and we want to make sure we are all working together to support this hectic activity. According to the original schedule, UAT for the clients was supposed to only last for a few days, but the clients got really worried that this wasn't enough time for them to review the entire site in-depth so they requested more time. They didn't want to change the cutover date, so we gave them some options for increasing their test time and they agreed to the following:

1. The clients will review each form/tool immediately after they're approved by the internal teams during SIT.

UAT Prep Meeting Agenda

- UAT Schedule
- Approach
- Roles and Responsibilities
- Dashboard Review
- Meeting Schedule
- Guidelines
- Next Steps

Figure 10.4   Meet internally before UAT begins to discuss all the key elements, ensuring everybody is on the same page

2. The clients will begin their content review at the same time that the creative team starts theirs, giving them an additional five days of review time but potentially exposing them to some errors.

Approach: When deciding the approach to UAT, we need input from the account team that is primarily responsible for managing the clients, the creative team that needs to provide content updates per the client requests, and the test team that is ultimately responsible to ensure the site is approved for cutover. Since the clients and creative agency are not local (they're in Arizona), we need to be very clear with our communications, but we're all working toward the same goal here, so we should be fine. After a little deliberation the team has come to the conclusion that the creative team will sit down in person with the clients for every UAT review, along with their account lead who will open the tickets for them. The process will be as shown in Figure 10.5 depending on whether the fix is content related or if it requires a developer. Remember—no new changes are allowed at this point. We're only correcting mistakes.

Roles and responsibilities: Within each of the boxes shown in Figure 10.5, there are various roles that need to be filled by one team or another. Here are a few bullet points to consider:

- During the client review, things can move pretty quickly. Will there be a person in the room opening tickets on the fly, or will we be logging them and then opening the tickets up later? Will more than one person be required to help with this since we need to log URLs, grab screen shots, and write down a description of the issue?
- Will there be somebody in the room that will be able to determine if the issue qualifies as a bug or a change? Do we need somebody from the tech side on the phone to help, in order to manage expectations?
- We should have one person at the creative agency and one at the production agency to whom we can assign all tickets. That person can then forward the ticket to the correct resource.
- Is the client responsible for closing tickets or should the agency be responsible for that task?

Dashboard review: Another thing we should discuss, and even review, is the dashboard for the ticketing system. Let's figure out how we want

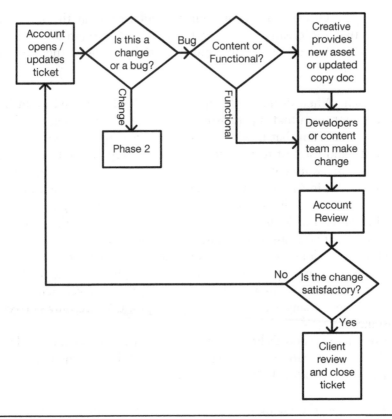

Figure 10.5   Take time to set up a process for UAT, so everybody is clear how the project will progress

the dashboard laid out and then come up with labels that will support the desired configuration. Figure 10.6 shows a really basic example for a dashboard, but of course, they can get quite complex. I always like to have a pie chart with the names of people who currently have tickets assigned to them, along with the number of tickets in their queue. That way I can see that Tim, who's a developer, has a relatively high number (11) of tickets in his queue, so I might want to contact him to make sure he knows what the deadline is for getting them done and ask if he has a conflict. Or, I could pull up the client's tickets to make sure they understand what their next action should be on each of them. The pie chart doesn't require a special label because the ticketing system can already make this query.

The other *gadgets* I have on the dashboard include a section for critical tickets and a section for tool-specific tickets. Critical tickets are

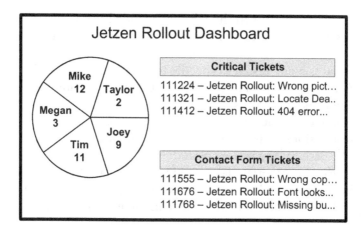

Figure 10.6   The ticket dashboard is an excellent tool for managing UAT status meetings

always top priority for us because if we don't get those fixed, it may delay launch. Since *issue type* is a required attribute in each ticket, we probably don't need a label for this either—a good bug tracking system will allow users to create a varied dashboard, based on all available inputs or data elements. But if we want to track each tool independently, then we do need a label—because nowhere in the ticket do we indicate through a drop down or open field if the issue is connected to one of our tools. We'll come up with a label for each tool we want tracked and then make sure everybody has that information before UAT begins, so they can enter them as they go.

Meeting schedule: During UAT we may have to increase our status meeting cadence and have them occur more frequently. I don't need to over explain this one, but there's a lot going on right now and cutover is just around the corner. Maybe the content team needs to meet every day at 4:30 to check and see if there are so many tickets that overtime is required. Or, perhaps the developers want to have stand-ups every morning to make sure they understand what the critical issues are for the day. The client may even want to meet daily or every other day to review the dashboard to make sure that things are getting done. It's all hands on deck right now and we need everyone to focus.

Guidelines: By guidelines I mean more of an instructional sheet for people to follow when conducting their own testing. Sort of like the test scripts that had all of the pertinent information in one place for people to follow along with. Even though the extended team won't follow certain steps, they'll need to know the following information:

- What is being tested (a specific tool or sprint package?)
- A link to the testing environment along with any required user name and passwords
- A link to the ticketing system and dashboard
- Instructions on how to enter tickets and input labels
- The dates of the review period
- A list of appropriate labels and when they should be used
- Instructions on how to close out tickets, who to forward them to, and the protocol for providing comments or resolving issues

Next steps: For next steps we should review the rest of the schedule and even start to discuss the cutover plan, which I'll talk about in the next chapter. There are a lot of steps that need to happen prior to cutover, and they should be in progress at this time. Other than that we just have to schedule the UAT review meetings, get the dashboard set up, and roll it out.

NONFUNCTIONAL TESTING

At the same time we're conducting UAT, the nonfunctional testing should be taking place in a separate, yet similar, environment. Remember that the nonfunctional tests include performance, security, disaster, and failover testing. As reported in Chapter 8, we have to wait until about the time that UAT begins before we can start the nonfunctional tests, because we need to be testing a version closest to what will be going live. If we test an incomplete site, then our tests won't be valid. But this leaves us with a bit of a problem, doesn't it? What if we don't pass some of these nonfunctional tests? It seems like pretty late in the game to be finding that out when we're about to launch this project. Well, let's play this through. I'll list each of the nonfunctional test types, add a *what-if* scenario, and try to figure out what would really happen if our site failed any of these tests.

Performance—what if our configurator pages weren't loading fast enough due to the large amount of calls it has to make to build the customer's perfect spaceship? Option 1 for each of these *what-if* scenarios is that we don't launch until the problem is solved. But, there's usually a workaround of some sort. For some of them, like the page load issue, the workaround is to just let it slide and try to fix it later. Page load times aren't really a reason to delay a launch, and I can tell you from experience that if the configurator is experiencing longer load times due to the number of data and image calls it has to make, the fix isn't going to be simple. We'd have to rebuild the whole design, and possibly the database systems that are supporting the tool along with it. These are risks that should have been considered during the content strategy discussion. But, sometimes we start with something quite simple, and after everybody and their cousin give their input and suggestions for an enhanced experience, we end up making things so complex even the customers give up on it.

Security—what if we found out that the contact form wasn't completely secure, leaving our customer's data vulnerable? This is something completely unacceptable and we can't launch this page. We should just turn off this particular page until it's fixed.

Disaster recovery—what if this test just downright failed and we found out that the backup system wasn't as stable as we thought? That's scary—I mean really scary—we could lose everything. I don't think delaying launch would either fix it or create more of a risk, but we should definitely not delete the *old* site until the problem is fixed.

Failover—what if we found out that our redundant hardware breaks down just as quickly as the equipment it's supposed to be covering for? This is the same situation as disaster recovery. There's no reason to delay the launch, but order a new piece of equipment as soon as possible. This is the reason we conduct these tests—to find out if the equipment is still working—just like we do with our smoke alarms at home. There's no reason to stop cooking dinner, but get some new batteries, eh?

See, we always have options. At the end of all this testing there may be a list of things that the client has requested to be fixed or changed but that we weren't able to get to before launch. Just be really clear with the team about which fixes are still outstanding and by what date we expect those fixes to be implemented. And now that we've cleared all of our tests, it's time for the big day. Cutover! T-minus 5, 4, 3, 2...

SOMETHING EXTRA—GLOBAL ROLLOUTS

You know what's really, really, really fun?—I mean besides waterslides?— building and consolidating several different websites for a global brand that currently has several different agencies around the world who are building several different websites experiences for them, and consolidating both the agencies and the websites in order to send one brand message globally. This is especially so when we're using a content management system to template the design, eliminate wasteful spending, and implement a governance system. It's my favorite type of project. One reason I love it so much is that it really does make sense for the brand and the company's financials.

We know from Chapter 1 that Jetzen has global offices in Mannheim, Germany and Shanghai, China. Before the *global economy* came about many years ago, Jetzen probably hired local ad agencies close to each of their three global headquarters because of many issues, including language barriers and local marketing strategies. And even though Jetzen had a brand styleguide for everybody to follow, there was just no way around having each client and each creative agency putting their own individual flair into their campaigns. Technology has reached a place where a brand can change a message in one place and have that exact same message sent out around the world immediately to every computer, cell phone, and tablet; via text, e-mail, or browser.

Figure 10.7 shows that global rollouts follow very much the same process as we follow for single sites, we just need to add a localization phase. To complete this global rollout for Jetzen's three sites, we'll be working with the global headquarters in Arizona for the initiation and planning phases, and then with the offices in Mannheim, Germany and Shanghai, China for the localization and Construct to Close phases. It's really not that complicated, I just want to point out a few main differences.

Initiation is basically the same as any other assignment. Before we get too far into any project, we need to complete gap analysis and produce some documentation against that assessment, and attach a budget. There are a couple of things we need to add into the mix, though. First, we'll need to include the other two countries in on our gap analysis so that we'll have a basic understanding of Jetzen's global business model before making any recommendations. Also, we'll want to make sure that the Arizona clients understand that once we start talking to the folks in Mannheim and Shanghai, they may have some additional requirements that we're not aware of yet, and new requirements will cost more money. Jetzen will have to work out who ends up paying for those additional requirements.

Communication planning: The communication plan will probably change a bit if you need to include global clients. The Arizona clients have to help us determine what information they're comfortable sharing with Mannheim and Shanghai, and what types of conversations should be discussed just between us and the corporate office. Several different meetings and status reports may end up being part of our overall communication plan.

Sitemap: The strategy here should be to make one sitemap that will work for all the countries. Of course there may be some differences by country—they may have different pricing, different warranty or other legal obligations, and different events and so forth, but for the most part we should be able to

Figure 10.7 Global rollouts have a localization phase for language translations and country-specific adjustments to marketing campaigns, promotional offers, and pricing

make something that everybody can use. Having a consistent global sitemap will set the stage for having a global brand that provides the same customer experience no matter what language is used.

Mobile assessment: The purpose of the mobile assessment is to compare Internet bandwidth capabilities and major mobile providers of the other locations to the U.S. to see if there is a gap in the available services, equipment, or capacity. With the two countries we're dealing with—Germany and China—there shouldn't be a problem, as they should be pretty much equal to or ahead of the U.S., but if we were dealing with smaller markets, they might have some gaps that would affect our content strategy.

Test strategy: Although the test strategy should be the same, I took the test cases and dropped them into the localization phase, since every website should have different instructions. Sure, they can share quite a bit, but each one will have to be customized at least to reflect the different domain names for each country.

Content localization and translations: Translations are an obvious requirement for copy shown on the site, but there's a lot more to factor in regarding localization. Although we're going to build a sitemap and wireframes that work for everybody, the local clients should have a chance to explain any cultural or technical differences that require adaptability. I suggest conducting in-market workshops and going through every aspect of the website with the local teams—because there are *many* things that might come up here. Here are just a few examples:

- If the spaceships are manufactured at different plants, then there's a chance that the end products will differ from what's sold in the U.S. China may have different government or legal requirements to account for, and once these are added into their production run the spaceships may actually look different. Watch out for this if they want to share photography with the U.S. or Germany.
- Marketing strategies could vary widely, even if the spaceships are absolutely the same. The U.S. could be marketing Atlantic Star as a weekend getaway vehicle, while China has had such a traffic problem on the ground that they've already set up atmospheric infrastructure to allow for using these spaceships to travel to and from work. The local countries should be able to alter campaigns to reflect what sells best in their markets.
- The travel path for *bit* may be different for these countries and require a customized technical solution strategy. Their corporate structure may also require that the forms and tools be altered to reflect the way they do business compared to the U.S.

Change management: Throughout the localization conversations and assessment, the change management process will have to be implemented. Make sure to talk to global headquarters before setting sail to talk with the local markets, so that it's understood what decisions the local markets are allowed to make and what decisions need to be approved by corporate. It's usually based on who's paying for the change.

Construct to Close: Not too much changes in the Construct to Close process, but of course, we'll be working with several different clients, agencies, and third parties to get these sites built and launched. Probably the content strategy will be altered to reflect content sharing between countries, but all the major steps still need to be conducted per normal. I highly recommend not trying to launch all three countries at the same time, unless the production and infrastructure teams are quite large. Stagger everything like we did in our agile approach to each task in the project plan. Get the U.S. off and running, then go to Germany and get them off and running, and then off to China. Put at least a week, if not more, between cutovers to ensure that the sites are stable after they launch, before shifting focus to another country's cutover.

In this chapter we:

- ✔ Completed systems integration testing
- ✔ Ran through end-to-end testing
- ✔ Held a UAT prep meeting and then conducted UAT
- ✔ Conducted the nonfunctional testing

11

CUTOVER

When people talk about *cutover* they are referring to the date and time that the new code is made available to the public or target audience. Other words used to capture this big moment are *launch, release,* or *going live.* If the team has a solid cutover plan in place, there shouldn't be any cause to worry. In this chapter we'll focus on learning the cutover process that I've developed through years of trial and research, and although we are applying it to the launch of a new website, it can be applied even if we're simply incorporating new tools into an existing website.

I bet it won't be surprising to hear that the first time I managed a cutover all by myself, there weren't any instructions or documentation to follow. I knew these two things for sure: first, if the cutover wasn't managed well and ultimately failed, then my job would probably be in jeopardy; and second, I knew that my team had a full year or so of various cutovers ahead of us, and that it was my responsibility to teach the other project managers (PMs) how to manage this important succession of tasks successfully. We needed a bullet-proof, *step-by-step process* for everyone to use, because most of the people on my team didn't have any experience with cutovers either.

Based on these two things, I decided to create a standard cutover plan that has since been successfully utilized over and over again by multiple PMs. Even the infrastructure teams and technical directors whom I work with refer to it when other PMs inevitably end up in their offices with the same confused look that I had a few years ago. The cutover plan is an Excel spreadsheet with three tabs. The first tab lists the steps to take *before* cutover day, the second tab lists the steps to take *on* cutover day, and the third tab lists all of the stakeholders we've been keeping track of since that first workshop we held with the Jetzen team months ago. It's very specific and detailed because just a tiny oversight during this process could ruin the cutover, so I suggest making a solid plan and then following it closely.

301 REDIRECTS

Before I get too deep into the cutover plan, I want to discuss 301 re-directs exclusively. They're critical to understand and can sometimes be such a large effort, that we'll want to start working with the search engine optimization (SEO) team and planning things out well in advance of when we'll actually need the list of redirects. Once I take you through what 301 redirects are, it will be easier to understand why the SEO team will require so much time putting them together, and then once we review the cutover plan together, I can show where they fall into place.

Earlier in the book, during initiation, we learned that Jetzen has an existing website and that our job was to help with the redesign. This included building them a whole new site because their existing site is very outdated and in need of some major remodeling. We got rid of most of their current pages and added in a bunch more. What that means for us and for our cutover is that Jetzen has a list of URLs out there that will no longer be in use after the cutover (the deleted pages) and a new list of URLs that we want customers to easily find (the new pages). The deleted pages are the problem. Let's look at an example.

Figure 11.1 shows Jetzen's original sitemap. Notice that there are four pages crossed out. *Our Products* is crossed out because we now have a landing page called *Model Selector*, which is quite similar but functions a bit differently. *Photos* and *Brochures* are crossed out because, instead of standing up single pages to house all of the photos and brochures, we're now including them within each model's features pages. And the Blog just won't exist any longer. So the problem is that there are URLs out there that will no longer apply once we cutover to the new site. We don't want anybody trying to find *www.jetzen.com/blog* after cutover be-cause it won't exist—we instead want them to find their way to a new

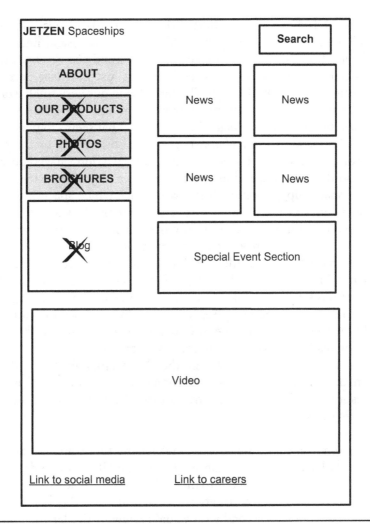

Figure 11.1 Jetzen's sitemap for their current live site, showing some pages we've deleted in the new design

page that might contain the same or similar content. So what the SEO team needs to do is to create a list of all the URLs that will no longer exist within the new site, and then map them to one of the new pages. And that's why they're called *301s*. In http terminology, the code 301 indicates that the page has been permanently moved to another URL.

Table 11.1 shows a very short 301 redirect list—depending on the size of the project, the 301 list could be over a thousand lines. When producing this list, I always make sure to remind the SEO team that they need to give us the entire address, not just the */blog* part. Each address needs to start with *http*. The first row in our very small example shows that the SEO team wants to point what used to be the *Our Products* page to Jetzen's new *Model Selector* page. That seems to make sense, right? The *Photos* and *Brochures* pages are also pointing to the Model Selector page. Does that seem strange? Well, we don't have equivalent pages on the new site, and in order to get to a brochure or to see photos of a spaceship, customers first have to pick which spaceship they're interested in. On the existing site, Jetzen put the brochures for all of the spaceships on one page, but now they each live on their respective features pages. So, it makes sense to me that we'd send people to the Selector page first. The other option might be to direct them to the homepage. Finally, for the blog page that will no longer exist at all, the SEO team is sending them to the News Stories landing page. Again, I'd think that the only other option there would be to send customers to the homepage, but maybe the SEO team has some research that shows that blog readers are going there for news content.

Once I get the final 301 list from the SEO team, I like to run through it to make sure it's perfect before sending it to the infrastructure team for testing. I've learned a few tricks over the years—and now you'll know them too:

1. **Remove duplicate URLs:** Make sure there are no duplicate *old* URLs listed. We can only send the URL to one destination page—we can't send it to multiple pages. Look for any duplicates by sorting the information in column A of the Excel spreadsheet, and then conduct a visual review of the URLs. Although Excel provides the ability to automatically remove duplicate URLs, I don't like using this feature because it just removes the

Table 11.1   301 Redirect list

OLD URLS	NEW URLS
http://www.jetzen.com/our-products	http://www.jetzen.com/model-selector
http://www.jetzen.com/photos	http://www.jetzen.com/model-selector
http://www.jetzen.com/brochures	http://www.jetzen.com/model-selector
http://www.jetzen.com/blog	http://www.jetzen.com/news

duplicates without letting me know what was deleted. The list would be fixed—but I like to point out the discrepancies to the SEO team and ask them to select the preferred destination.

2. **Remove special characters:** Sometimes when the SEO team puts together the existing URL list, they use web services that allow the user to drop in a domain name and then the system will provide a list of every single URL associated with that domain. Sometimes we get some funky URLs with special characters in them. It might look something like this: *www.jetzen. com/$...#_blog*. This happens all the time and we need to look out for those. Just send those right back to the SEO team for cleanup.

3. **Remove external links:** Another thing I always see are off-brand domain names popping up in the list. The way a lot of these sites are built, a user could be clicking around from page to page and not even realize that they've actually left the Jetzen site and are now looking at a *microsite* built by a third party. We do this all the time. Let's say Jetzen pays a third party to manage all of their parts supplies, and a big part of the service they offer is to warehouse and ship parts to both customers and service centers. To help customers find and purchase the part they're looking for, this company has built a special website that has various lookup methods customers can use to search for the part they need. Jetzen already has this web tool built and they're not going to pay us to build it all over again. So what we do is to send this other company our Cascading Style Sheet (CSS) information and even some of our code so that they can make their site look exactly like ours. It will even mimic our global header so customers can easily *get back* to our site—even though they probably won't even realize that they left in the first place. The only way the customer would notice that they're now on what we call a *microsite* is if they notice that the URL changed. The domain would be different—it couldn't be *Jetzen.com*—it would be something like *partsguy.com*. And to bring this lengthy discussion all the way back around to why I started talking about it in the first place, we have to remove all non-Jetzen website URLs from the list because we can only redirect domains that Jetzen owns and controls.

4. **Look for blank fields:** This seems pretty funny to mention, but I always find blank fields too. Usually I find some of the destination URL fields blank and I imagine that it's because the SEO

team wasn't sure where to send that particular page so they skipped it and forgot to go back and fill it in.

Jetzen doesn't currently have a mobile site, but if they did we'd have to redirect all of those pages as well. When companies don't build responsive websites and instead have one set of code for their desktop site and another set of code for their mobile site, the mobile pages would start with m. instead of www. If this were the case for Jetzen, we'd have to redirect every single mobile m. URL to one of the new www. URLs since none of the m. pages will exist after cutover. If you have to go through this process, be sure to create two separate 301 redirect lists and apply the same tricks I've explained above.

Now that we've got the final 301 redirect lists in-hand we'll just hold on to them until the cutover plan is approved, because we'll need to follow a specific process for testing and implementing them.

Cutover Management

Because there's not a lot of room to work with in book format, I'm going to just list the steps we need to follow for the pre-cutover activities, but note that in the real Excel file there's additional columns to list the date, owner, and comments for each activity. (Remember, the project files from the book are available from the J. Ross Publishing Web Added Value™ Download Resource Center at www.jrosspub.com.) The date column should be pre-populated with the dates by which these activities need to take place; it's not a column to list when the activities end up taking place. As I stated earlier, this is a very precise process that needs to be monitored and controlled on a daily basis. Table 11.2 shows the pre-cutover steps for Jetzen's launch. As always, I'm going to go through them all, to make sure everybody understands each activity.

Pre-Cutover Steps

Schedule nonfunctional testing: When host servers are managing many different websites they routinely schedule nonfunctional tests to take place each month. It's important to communicate well in advance with these folks, that we'll be adding a new site to their test schedule, and ask for instructions on how to ensure that the new site will be fully tested prior to launch.

Table 11.2 Pre-cutover steps

Jetzen Pre-Cutover Steps
Activity
Schedule nonfunctional testing
Conduct team meeting
Confirm secure servers
Send out cutover meeting notice
Prepare cutover schedule
Client approves cutover schedule
Test 301 redirects in pre-production
Request CDN change
Test CDN change in pre-production
Change DNS records
Check DNS servers
Request TTL to 10 minutes
Ensure SEO has been approved
Ensure Analytics has been approved
Test domain mapping in pre-production
Create master cutover ticket - Domain mapping (Ticket #123) - Remove authentication (Ticket #234) - CDN Change in Production (Ticket #345) - 301s (Ticket #456) - Purge CDN cache (Ticket #567)
Go/No-Go Decision

Conduct team meeting: This meeting should be held with everybody who has input to, or ownership of, the cutover activities and schedule. Every cutover presents its own set of challenges, so we can't just pick up the last plan we used and run with it. It's important to get everybody on a conference call and run through each activity that needs to be performed, both pre-cutover and on cutover day. Make sure the key people agree with our process and timing. Another reason to have this call is to make sure everybody knows that a major launch is coming up soon and that their support is crucial to its success—if someone isn't available to join in on the conference call, then track them down.

Confirm secure servers: Make sure that the people who manage the secure servers and those who pay for them are on the conference call to confirm that the servers are in place and ready to go. Once I was told that the client I was working with didn't have any secure servers available—it took them quite a while to procure the budget for, and then obtain the servers. We ended up launching the site without any forms, and then held a second cutover when they were finally ready.

Send out cutover meeting notice: Send out a meeting notice for the day of cutover so that everybody has it on their schedules, and track the invites to make sure that all the key players have accepted the meeting notice. I've found that sending out an actual meeting notice for this event often brings up resource conflicts—which are good to know in advance. Just hearing the date may not be enough to remind someone that there's another large activity planned for the same day, and if that happens there might not be enough resource coverage available for both activities.

Prepare cutover schedule: Once the team meeting has been conducted, we should have a pretty good process developed for both pre-cutover and for the day of cutover. Send this plan out to all the key parties for review and make sure that the client provides their approval. Remember, as these tasks are completed we can mark that down in the status column so that other team members can easily see where we're at in the overall process. Plus, marking down the word *complete* makes dopamine start blasting away in our PM heads.

Test 301 redirects in preproduction: Everything that goes into production must first be tested in a lower environment, and this is a good opportunity to make sure all of the little tricks we used to clear out unacceptable URLs or fields were successful. Some infrastructure managers will be really picky about this, and some won't. If we somehow missed a bad URL during our own review of the 301 list and it didn't get caught until after the infrastructure team had released the redirects into preproduction, then the correct way of handling it would be to remove everything from preproduction (roll it back). This is what I'm calling the picky way, but I agree—it's actually the right way to do it. We'd have to roll back our 301s, fix the list, and then resubmit it. The non-picky way would be if the infrastructure manager had the bad URLs removed and then replaced with the correct ones without making us roll everything back. The problem with the non-picky way is that the original list sometimes doesn't get updated appropriately and then bad URLs are accidentally put into production, which isn't good. The infrastructure manager is supposed to make sure that the *exact* same file we put into production was first tested in preproduction. Not simply the same list *after we made some fixes to it.*

Request CDN change: Work with the content delivery network (CDN) contact to get the code they'll be using for the new site. This is the code that tells the network where to find the new content (where the new servers are located) and where the content will be saved within the CDN for caching. Accepting this code into our system allows the two systems to communicate and share information with each other.

Of course, have this tested in preproduction, before it gets put into production.

Change DNS record: We need to make sure that the domain name system (DNS) record (www.jetzen.com) is pointing to the appropriate internet protocol (IP) address, where the content is stored. The old site would have been pointing to a different IP address, so we need to give them our new address.

Check DNS servers: After we've changed the DNS record, we just need to test that new connection.

Request TTL to 10 minutes: TTL stands for Time to Live, and means that once the new content is published or made available by our servers, it won't take more than 10 minutes for the content to be distributed worldwide through the content delivery network. We just need to know what the client's standards are for TTL and communicate this to the appropriate parties.

Ensure SEO has been approved: The publisher links for preproduction should be sent to the SEO team so they can validate that all of their recommendations have been implemented successfully.

Ensure analytics has been approved: The publisher links should also be sent to the analytics team so they can validate that all of their analytics recommendations have been implemented successfully. This includes them logging into their report suite and running the same reports they plan on running for the clients after launch.

Test domain mapping in preproduction: Now that we have the CDN properly mapped and connected to our IP address, we'll need to pull that connection down to the exact file directory.

Create all cutover tickets: Opening activity-specific tickets is an excellent way of ensuring that the key activity owners for cutover day have, and understand, their assignments. About a week before cutover day we're going to go into our ticketing system and create something called a *Master Cutover Ticket*. Under that ticket, we'll open the following sub-tickets, which represent every major activity that will be taking place on cutover day:

- Domain mapping (Ticket #123)
- Remove authentication (Ticket #234)
- CDN Change in Production (Ticket #345)
- 301s (Ticket #456)
- Purge CDN cache (Ticket #567)

Go/no-go decision: Once the master cutover ticket is in place and all the appropriate parties have approved those activities to take place on August 6, we'll be ready to go. Now we just need the business client to give us an official approval to launch. They'll review the last set of outstanding fixes—that we've already pointed out won't get fixed until after cutover—and give us the official decision.

Finally, most teams want to also include "back out" procedures just in case we're not able to get the site stable enough to launch. The back out procedures are simply doing the cutover steps in reverse. It's a real let down when the team has spent the past several months preparing for this critical day, and then things go so horribly wrong that we need to roll back the cutover. I hope this never happens to you!

Cutover Day

It's finally cutover day! We've waited all through the book for this—and now it's finally here! I hope you got some sleep last night because it's going to be a long day! But soon we can sleep ☺! Table 11.3 shows our cutover schedule which has been reviewed and approved by all parties. As shown in the example, we list each of the activities, the ticket numbers associated with the technical steps, start times, end times, durations, and owners. Notice how exact the time schedule is. If something just takes five minutes, we put it in there to make sure it's accounted for. I do this for two reasons:

1. If something is taking a lot longer than we expected, then I start to get nervous and question the activity owner. Is something wrong? Do I need to worry?
2. For those folks who don't want to be on the phone all morning, like perhaps the quality assurance (QA) team who don't start their sanity checks until about 10:45, we can give them an approximate time frame for when we'll need them on the call, so they can do other work in the meantime.

That first step of implementing the new domain mapping can take a long time, and that's because our package is part of a larger monthly release package with fixes and new code for everything in this particular production environment. Of course this doesn't happen every time, but make sure the infrastructure team lets you know how long that first

Table 11.3 Cutover schedule

Jetzen Cutover Steps					
Activity	**Ticket**	**Start Time**	**End Time**	**Duration**	**Owner**
Implement domain mapping	Tick-123	6am	9am	:180	Inf.
Call begins		9am	9:05	:05	Agency
Remove authentication	Tick-234	9:05	9:10	:10	Host Provider
Implement CDN change	Tick-345	9:10	10:10	:60	Inf.
Confirm DNS change is live		10:10	10:25	:15	CDN
Implement 301s	Tick-456	10:25	10:40	:15	Inf.
Purge CDN cache	Tick-567	10:40	10:45	:05	Inf.
Perform sanity check and remind participants to clear browser cache		10:45	11:45	:60	QA All
End-to-end testing		10:45	11:45	:60	QA All
Read out all reported findings		11:45	12pm	:15	Agency
Stay-alive/back-out decision		12pm	12pm	:00	Client

step will take, because *it's a doozie*. I don't even open up the conference line until I'm told that the code release is complete. Important to note is that between the time the release is complete and when we're done with that last ticket for clearing the CDN cache, the Jetzen site will be in disarray. Make sure everybody knows this or they will *freak out*—seriously! I tell the clients that cutovers entail several hours of instability and possibly downtime for their website as we switch over from one site to the other. Because of this, many clients choose to conduct their cutover in the middle of the night when traffic on the site is at its lowest—and this should be accommodated.

When the release is complete, I open up a conference line that will be kept open throughout the cutover. Ticket #234 shows that the first thing we do after the code is dropped into production is to have our host provider remove authentication. When we were just building the site our content was under lockdown, and testers had to submit a username and password in order to see it—but now we want the content to be open to the public so we're removing that authentication step. Next we'll have the CDN change implemented, and once that's done we'll check the DNS servers to make sure the change is live. Most of these activities will be owned by the infrastructure manager, but they sometimes have us put their various contacts at the CDN or host provider right into the cutover schedule as key contacts. Whichever the case, the infrastructure team is responsible for most of these activities and will be our first phone call if something goes wrong at any point during the day.

Now it's time to implement those 301 redirects that were such a hassle getting approved in preproduction. At this time in the process the sites are really a mess, so make sure nobody tries to open up www.jetzen.com until we give them the go-ahead. People will, though, they can't help it. And they may get a little anxious and report that the site is a mess, but just remain calm and ask the participants to refrain from testing until we reach that part of the schedule. I always tell them, "It's expected that the site will be unstable at this time, so what you're seeing is completely normal. Once we get through the cutover steps, I'll let the team know when they can start testing, which is expected to begin around 10:45." Or I say, "Stop freaking out—I told you that this would happen. Just chill until testing officially begins." It depends on whom I'm talking to.

We couldn't put the 301s into production until after our site was ready because if we had done it yesterday, the old URLs would be redirected to new URLs which weren't live yet. So a lot of 404 errors would appear. By the way, the Internet has a lot of standard error messages in place for this sort of problem. We also have the ability to publish our own error messages as long as they're associated with the appropriate codes and conditions. There are a ton of codes out there, but here's a list of the ones we see most often:

- **403 Forbidden:** This means that the website we're trying to look at is telling us, "Hit the road and don't come back" or, "You're not wanted here!" The most common reason for getting a 403 is that the user has dropped in an incorrect URL, but it also happens when trying to reach restricted web pages that require special access permissions.
- **404 Not found:** This is a nicer way of telling us, "Hmm, this URL isn't showing anything, are you sure you have the right address?" Either the URL is wrong or the page isn't publishing for some reason. Like in the case of having 301 redirects sending users to pages that haven't been published yet—the URL is right, but it's just not working right now.
- **502 Bad gateway:** I see this one a lot too. It means that something is going wrong between the servers and the browsers, but I have no idea how to fix it.

404 is usually the only error message that I can attempt to look into and figure out for myself, but for most error messages, we should go to the infrastructure manager and let them handle it. Anyway, back to the cutover, the 301s have been implemented and we're about ready to begin sanity checks. The only thing left to do is to clear our caches. Both the CDN cache and our personal browser caches need to be cleared out. This will erase any content still hanging around from the old site and ensure that just the new stuff is being published. We'll give folks about an hour or so to check the site, clicking on everything, opening videos, and just doing our best to find something that's not working right. Also during this time, we can get the end-to-end team on the line to make sure all the applications and forms are working as expected.

Throughout the process, encourage people to either e-mail the errors they're finding or enter them into the ticketing system. They'll want to just shout them out over the conference line, but that method can get pretty hectic and unmanageable if there are a lot of suspected errors found. Many of them can be fixed on the spot—especially if we're dealing with a content management system instead of an html site. For things we can't easily fix, simply open a ticket and assign it to the appropriate resource. After an hour or so we'll ask everybody if they've completed their checks and then call out the known errors. Finally, we ask the client if we have permission to end the call and stay live, because otherwise we have to do a rollback.

In all the cutovers I've been associated with, I *have* been in the situation where we've decided to delay launch by a day or a week or so because we knew it wasn't ready, but I've never been in the middle of a cutover and then had things go so terribly wrong that we had to rollback and go back to the old site. I've seen people get really close to this, but nobody has ever pulled the plug. We just stay up, keep the line open, and continue working on the issues until whatever was broken starts working again. If we did have to roll back the site we'd refer to the *back-out* process we established, just in case of an emergency. It basically involves taking the cutover steps and flipping them upside down—doing them in reverse. Beep, beep, beep—we're backing up—watch out! That's really quite a failure on somebody's part, and I don't wish that on any readers of this book!

TRANSITION TO OPERATIONS

After launch, the operations team may request that the build team stay on point for one to two weeks after cutover to handle any outstanding issues and to ensure the site is stabilized. The build team will want to take a very long nap and drop out of society for a while, and we can take turns with that, but make sure somebody is around to answer questions as the brand or operations team start to take control. One important part of the transition to operations period is to prepare some training documentation. I've been on projects where the operations team has had very little participation with the cutover, and projects where the operations team was critical to our success. Either way, there should be some documentation prepared; but of course a team that has had little exposure to the process thus far will require much more training. I'm not going to explain everything that needs to be said during the transition to operations training, but Figure 11.2 shows a slide that lists some topics that should probably be covered.

One of the reasons our sustain team wouldn't have much knowledge about the project we just launched would be if they were from out of town. Like, *way* out of town. Many companies these days find offshore digital production teams to be an excellent way to save money. For

---

Transition to Operations   JETZEN
                                       SPACESHIPS

- Client Contact Information
- Extended Team Contact Information
- Sitemap Review
- Forms, Applications and Tools
- Process for Making Updates
- Styleguide

Figure 11.2 Transition to operations agenda

Jetzen, we're transitioning the site maintenance and sustain activities to an outstanding digital firm they've hired out of Costa Rica. So, knowing that the new team hasn't had any exposure to the website or even our clients, we need to make sure they're provided with a full overview of the project from start to finish. First, we're going to provide them with the client contact information. That is, unless we aren't supposed to. Maybe these guys work directly with the creative agency. Whatever the case, let them know who their contacts are.

Next, they need to know who all of the extended team members are, even if they're never supposed to speak with them directly. I once had a boss who wouldn't give me the whole picture on things—he'd always say, "You don't need to know that." But, I do! It's really hard to make intelligent decisions and recommendations when you don't know the whole story. So I always give people a thorough explanation of the current situation. Here are the clients, this is how we got involved with the project in the first place, these are the people who handle SEO, these are the analytics folks, and make sure you never talk to Larry about Dave because those guys can't stand each other.

The next several bullets on the slide should be quite clear. We'll want to review the sitemap with them, since it provides an outline of the entire site, go through all of the forms, applications and tools, and show them any process documentation we have regarding how changes will be submitted and how they're expected to respond. It's also a good idea to run through the styleguide since it provides information about the types of assets that will be coming in and what the general rules are supposed to be regarding staying on track with the design. If the team will be using a content management system to update the content, then they'll require several days of training on the templates and components, and developers will need their own training on how the code was developed. In short, show them everything. You may even plan on flying down there for at least a couple weeks to conduct this training and hold their hands as they start to take command of the ship.

Well, we're done. It was a really fun assignment, but after everything's considered, this project will go down in history as one of the biggest disappointments of my career since I never got to ride in a spaceship. But still, onward and upward for our careers, right?

In this chapter we:

✔ Created the 301 list
✔ Developed and applied all pre-cutover steps
✔ Launched www.jetzen.com

12

SUMMARY

"The rollout process was laid out for me when I started my first digital project manager job, and as the newest member of the team, my assignment was to create a training document outlining the steps. This allowed me to familiarize myself with it and now I use the rollout process for every single site project I manage. It pretty much ensures success thanks to the straightforward, intuitive instructions."

—Megan Reed, Rollout Manager

I think Megan was wondering what the heck she got herself into when she found herself learning process and writing training documents when we first started working together. I feel so proud when I see her and other project managers (PMs) whom I've taught the process to, applying it to their jobs long after they've left my team. You know, writing a book and claiming to be an expert on a subject can be quite scary—what if people disagree with me and try to discredit my process? Well, I'm not going to be able to please everybody, but I *know* that this process works. I've used it myself over and over again, and so have my colleagues around the globe. And honestly, I can't wait to hear from people through Twitter (@RolloutManager) or LinkedIn, and engage in conversations about other perspectives. It's all good. I just really hope it helps people.

In addition to the process, I want to stress the importance of using a professional rollout manager to lead digital efforts. The rollout manager specializes in the management of technical projects from gap analysis to content creation and cutover planning. They're responsible for coordinating all efforts of the workflow, from maintaining a detailed and up-to-date project plan, communication planning, and status reporting, to understanding the technical aspects and risks associated with every task and resource. The rollout manager is the direct contact for all offshore and cross-functional teams and vendors including analytics, search engine optimization, development, creative, and user experience. In short, the rollout PM is ultimately responsible for coordinating every aspect of the project. If your company doesn't have this resource on staff, you may want to start looking for one. Or, maybe *you're* going to be this person!?

Well, it's been a pleasure working with you. Wouldn't it be fun to see the Jetzen website up and live for real? Too bad it was all just a case study. I hope you've enjoyed the project and learning about each of the 30 steps. Take care, and good luck with your future rollouts!

APPENDIX A: COMPLETE PROJECT PLAN

	Task	Days	Start	Finish	Predecessor	Resource
1						
2	Gap Analysis	0	Mon 1/5/15	Mon 1/5/15		IA
3	BRD	0	Mon 1/26/15	Mon 1/26/15		IA,PM
4	Statement of Work	0	Fri 1/30/15	Fri 1/30/15		
5	**Sitemap**	**9**	**Mon 2/2/15**	**Thu 2/12/15**		
6	Sitemap Developed	5	Mon 2/2/15	Fri 2/6/15		IA
7	Review Sitemap	1	Mon 2/9/15	Mon 2/9/15	6	ALL
8	Make Changes	2	Tue 2/10/15	Wed 2/11/15	7	IA
9	Sitemap Approved	1	Thu 2/12/15	Thu 2/12/15	8	Client
10	**Group 1**	**25**	**Fri 2/13/15**	**Thu 3/19/15**		
11	Wireframe Development	5	Fri 2/13/15	Thu 2/19/15	9	IA
12	Wireframe Presentation	1	Fri 2/20/15	Fri 2/20/15	11	IA
13	Revisions Distributed	2	Mon 2/23/15	Tue 2/24/15	12	IA
14	Client approval	1	Wed 2/25/15	Wed 2/25/15	13	Client
15	Comp Development	10	Thu 2/26/15	Wed 3/11/15	14	Design
16	Styleguide Development	5	Thu 3/12/15	Wed 3/18/15	15	Creative
17	Styleguide Presentation	1	Thu 3/19/15	Thu 3/19/15	16	Creative
18	**Group 2**	**26**	**Thu 2/26/15**	**Thu 4/2/15**		
19	Wireframe Development	5	Thu 2/26/15	Wed 3/4/15	14	IA
20	Wireframe Presentation	1	Thu 3/5/15	Thu 3/5/15	19	IA
21	Revisions Distributed	2	Fri 3/6/15	Mon 3/9/15	20	IA
22	Client approval	1	Tue 3/10/15	Tue 3/10/15	21	Client
23	Comp Development	10	Thu 3/12/15	Wed 3/25/15	15	Design
24	Styleguide Development	5	Thu 3/26/15	Wed 4/1/15	23	Creative
25	Styleguide Presentation	1	Thu 4/2/15	Thu 4/2/15	24	Creative

26	Group 3	105	Wed 3/11/15	Tue 8/4/15		
27	Wireframe Development	5	Wed 3/11/15	Tue 3/17/15	22	IA
28	Wireframe Presentation	1	Wed 3/18/15	Wed 3/18/15	27	IA
29	Revisions Distributed	2	Thu 3/19/15	Fri 3/20/15	28	IA
30	Client Approval	1	Mon 3/23/15	Mon 3/23/15	29	Client
31	Comp Development	10	Thu 3/26/15	Thu 4/9/15		
32	Initial Creative Development	3	Thu 3/26/15	Mon 3/30/15	23	Design
33	Internal Review	1	Tue 3/31/15	Tue 3/31/15	32	
34	Comps Are Updated	0	Wed 4/1/15	Wed 4/1/15	33	Design
35	Client Review	1	Wed 4/1/15	Wed 4/1/15	34	Client
36	Copy Development	1	Thu 4/2/15	Thu 4/2/15	35	Copy
37	Comps Are Updated	1	Fri 4/3/15	Fri 4/3/15	36	Design
38	Proofreading Review	1	Mon 4/6/15	Mon 4/6/15	37	Proof
39	Final Client Review	1	Tue 4/7/15	Tue 4/7/15	38	Client
40	Legal Review	1	Wed 4/8/15	Wed 4/8/15	39	Legal
41	Delivery	0	Thu 4/9/15	Thu 4/9/15	40	Design
42	Styleguide Development	5	Thu 4/9/15	Wed 4/15/15		Creative
43	Styleguide Presentation	1	Thu 4/16/15	Thu 4/16/15	42	Creative
44	Group 4	28	Tue 3/24/15	Thu 4/30/15		
45	Wireframe Development	5	Tue 3/24/15	Mon 3/30/15	30	IA
46	Wireframe Presentation	1	Tue 3/31/15	Tue 3/31/15	45	IA
47	Revisions Distributed	2	Wed 4/1/15	Thu 4/2/15	46	IA
48	Client Approval	1	Fri 4/3/15	Fri 4/3/15	47	Client
49	Comp Development	10	Thu 4/9/15	Wed 4/22/15	31	Design
50	Styleguide Development	5	Thu 4/23/15	Wed 4/29/15	49	Creative
51	Styleguide Presentation	1	Thu 4/30/15	Thu 4/30/15	50	Creative

52	**Group 5**	**29**	**Mon 4/6/15**	**Thu 5/14/15**		
53	Wireframe Development	5	Mon 4/6/15	Fri 4/10/15	48	IA
54	Wireframe Presentation	1	Mon 4/13/15	Mon 4/13/15	53	IA
55	Revisions Distributed	2	Tue 4/14/15	Wed 4/15/15	54	IA
56	Client Approval	1	Thu 4/16/15	Thu 4/16/15	55	Client
57	Comp Development	10	Thu 4/23/15	Wed 5/6/15	49	Design
58	Styleguide Development	5	Thu 5/7/15	Wed 5/13/15	57	Creative
59	Styleguide Presentation	1	Thu 5/14/15	Thu 5/14/15	58	Creative
60	**Scope Freeze**	**1**	**Fri 5/15/15**	**Fri 5/15/15**	**59**	
61	**Analysis and Final Prep**	**13**	**Fri 5/15/15**	**Tue 6/2/15**		
62	Create Test Strategy/Cases	10	Fri 5/15/15	Thu 5/28/15	59	QA
63	Review Test Strategy/Cases	3	Fri 5/29/15	Tue 6/2/15	62	PM
64	Analytics Analysis	10	Fri 5/15/15	Thu 5/28/15	59	Analytics
65	SEO Analysis	10	Fri 5/15/15	Thu 5/28/15	59	SEO
66	**Development**	**82**	**Fri 3/20/15**	**Mon 7/13/15**		
67	CSS	50	Fri 3/20/15	Thu 5/28/15	17	FE Dev
68	HTML Framework	60	Fri 3/20/15	Thu 6/11/15	17	Web Dev
69	Search	4	Fri 3/20/15	Wed 3/25/15	17	Dev 1
70	News Stories	25	Fri 5/1/15	Thu 6/4/15	51	Dev 2
71	Video Gallery	14	Fri 4/17/15	Wed 5/6/15	43	Dev 1
72	Configurator	42	Fri 5/15/15	Mon 7/13/15	59	Dev 3
73	Trip Planner	26	Fri 6/5/15	Fri 7/10/15	71	Paul
74	Contact Us Form	12	Fri 5/15/15	Mon 6/1/15	59	Dev 4
75	**Asset Delivery**	**16**	**Mon 6/1/15**	**Mon 6/22/15**		
76	Group 1	1	Mon 6/1/15	Mon 6/1/15		Creative
77	Group 2	1	Mon 6/8/15	Mon 6/8/15		Creative
78	Group 3	1	Mon 6/22/15	Mon 6/22/15		Creative
79	**Internal AQR**	**17**	**Tue 6/2/15**	**Wed 6/24/15**		
80	Group 1	2	Tue 6/2/15	Wed 6/3/15	76	Asset Mgr
81	Group 2	2	Tue 6/9/15	Wed 6/10/15	77	Asset Mgr
82	Group 3	2	Tue 6/23/15	Wed 6/24/15	78	Asset Mgr
83	**AQR Fixes**	**16**	**Thu 6/4/15**	**Thu 6/25/15**		
84	Group 1	1	Thu 6/4/15	Thu 6/4/15	80	Creative
85	Group 2	1	Thu 6/11/15	Thu 6/11/15	81	Creative
86	Group 3	1	Thu 6/25/15	Thu 6/25/15	82	Creative

87	Content Input	20	Fri 6/5/15	Thu 7/2/15		
88	Group 1	5	Fri 6/5/15	Thu 6/11/15	84	Dev 1
89	Group 2	5	Fri 6/12/15	Thu 6/18/15	85	Dev 2
90	Group 3	5	Fri 6/26/15	Thu 7/2/15	86	Dev 3
91	**Internal QA**	**17**	**Fri 6/12/15**	**Mon 7/6/15**		
92	Group 1	2	Fri 6/12/15	Mon 6/15/15	88	QA
93	Group 2	2	Fri 6/19/15	Mon 6/22/15	89	QA
94	Group 3	2	Fri 7/3/15	Mon 7/6/15	90	QA
95	**SIT**	**84**	**Thu 3/26/15**	**Tue 7/21/15**		
96	**Search**	**5**	**Thu 3/26/15**	**Wed 4/1/15**		
97	Code in Pre-Prod	1	Thu 3/26/15	Thu 3/26/15	69	
98	Internal Review	2	Fri 3/27/15	Mon 3/30/15	97	
99	Code Fixes	2	Tue 3/31/15	Wed 4/1/15	98	
100	**Video Gallery**	**5**	**Thu 5/7/15**	**Wed 5/13/15**		
101	Code in Pre-Prod	1	Thu 5/7/15	Thu 5/7/15	71	Tim
102	Internal Review	2	Fri 5/8/15	Mon 5/11/15	101	QA
103	Code Fixes	2	Tue 5/12/15	Wed 5/13/15	102	Tim
104	**Contact Us**	**5**	**Thu 6/4/15**	**Wed 6/10/15**		
105	Code in Pre-Prod	1	Thu 6/4/15	Thu 6/4/15	74 FS+2	Tim
106	Internal Review	2	Fri 6/5/15	Mon 6/8/15	105	QA
107	Code Fixes	2	Tue 6/9/15	Wed 6/10/15	106	Tim
108	**News Stories**	**5**	**Thu 6/11/15**	**Wed 6/17/15**		
109	Code in Pre-Prod	1	Thu 6/11/15	Thu 6/11/15	70 FS+4	Paul
110	Internal Review	2	Fri 6/12/15	Mon 6/15/15	109	QA
111	Code Fixes	2	Tue 6/16/15	Wed 6/17/15	110	Paul
112	**Trip Planner**	**4**	**Thu 7/16/15**	**Tue 7/21/15**		
113	Code in Pre-Prod	0	Thu 7/16/15	Thu 7/16/15	73 FS+3	Paul
114	Internal Review	2	Thu 7/16/15	Fri 7/17/15	113	QA
115	Code Fixes	2	Mon 7/20/15	Tue 7/21/15	114	Paul
116	**Configurator**	**4**	**Thu 7/16/15**	**Tue 7/21/15**		
117	Code in Pre-Prod	1	Thu 7/16/15	Thu 7/16/15	72 FS+2	Charlie
118	Internal Review	1	Fri 7/17/15	Fri 7/17/15	117	QA
119	Code Fixes	2	Mon 7/20/15	Tue 7/21/15	118	Charlie
120	**UAT**	**50**	**Fri 5/29/15**	**Thu 8/6/15**		
121	Search	3	Fri 5/29/15	Tue 6/2/15	67	Team
122	Video Gallery	3	Fri 5/29/15	Tue 6/2/15	67	Team
123	Contact Us	3	Thu 6/11/15	Mon 6/15/15	107	Team
124	News Stories	3	Thu 6/18/15	Mon 6/22/15	111	Team
125	Trip Planner	3	Wed 7/22/15	Fri 7/24/15	115	Team
126	Creative Review	3	Wed 7/22/15	Fri 7/24/15	119	Creative
127	Fixes	3	Wed 7/29/15	Fri 7/31/15	126	Dev
128	Final Team Review	3	Mon 8/3/15	Wed 8/5/15	127	Team
129	Fixes	1	Thu 8/6/15	Thu 8/6/15	128	Dev
130	**Cutover**	1	Thu 8/6/15	Thu 8/6/15	129	Inf

APPENDIX B: GLOSSARY

301 redirect: In http terminology, the code 301 indicates that a page has been permanently moved to another URL. When implementing a 301 redirect for a website, it means that we're pointing an existing URL to a new one so that original page content is no longer available and instead the content from the new page is shown.

A/B testing: An experiment to test how a target audience reacts to two different designs. The *A* and *B* versions are both published separately for a set amount of time in the same environment, to see which one yields better results.

Acceptance criteria: Test results that need to be recorded before a unit is officially approved by the test team.

Alt tags: Used to increase search results, *alternative* tags are hidden captions behind images that can only be seen by search engines, screen readers, or while a user hovers over an image.

Analytics: The study of how users interact with a website or other digital project. This information is used to provide business owners with data they can study and track in order to design experiences that promote higher KPI (key performance indicator) results.

Architecture diagram: A graphic illustration of the data architecture requirements for a web tool or application used to help the tech team develop the correct solution.

As-is version: The current version of a digital solution that will soon be improved.

Asset delivery: The result of an asset provider packaging up and handing over content to another party.

Asset manager: A resource who tracks and monitors asset delivery and other related activities.

Asset quality review: Normally performed by an asset manager or content lead, this is the task of reviewing delivered assets against their approved specifications to ensure they match.

Assets: Any form of content to be used in the development of a digital project, including but not limited to images, copy, PDFs, or videos.

Automated testing: Using software to run repeatable test scripts instead of completing them manually.

Bandwidth: The rate at which data can be transmitted through a channel.

Banner ads: Named *banner ads* because the original designs literally looked like little banners waving in the wind, these are digital advertisements placed on and around the internet that allow click-through to destination websites.

Brand project manager: This resource is assigned to one particular brand full time, and knows the clients and their business inside and out.

Breadcrumb navigation: A graphic representation showing where users are within a site experience depicted by leaving a *trail* of pages visited.

Breakpoint: A responsive design technique which identifies the point at which columns within a grid system are either added or deleted based on the user's viewpoint.

Broadcast: Traditional radio and television advertising.

Broken links: This refers to hyperlinked text or buttons that are not providing click-through service because the hyperlinks are not connected to a destination page.

Bugs: Defects within the tool or application that is being tested.

Business analyst (BA): Requirements documenters who are skilled at taking technical concepts and explaining them to nontechnical teams in ways that are easily understood.

Business lead: Also called the *client,* this is the person who owns the project's budget allocation and therefore provides direction and makes all final decisions.

Business objectives: The desired outcomes of the project which are determined by the business owner.

Business requirements document (BRD): Provides detailed definitions of the requirements for each piece of development work so that the team has a thorough understanding of what exactly needs to be delivered.

Cache: Saved memory of visited pages stored in a browser or delivery network which is used to speed up content delivery and reduce strain on host servers.

Cascading Style Sheet (CSS): HTML language that provides the code for fonts, colors, and layout in one central location that's applied across the site.

Change control log: A spreadsheet that lists every single change request that comes in after scope freeze has been called. The log may contain information such as the name of the requestor, a description of the change, the date it was evaluated, the outcome of the evaluation, or costs associated with the change.

Change control process: The process the team follows to manage change requests.

Communication plan: A list of every communication requirement associated with a project, along with how they will be managed.

Comps: This refers to the design that art directors create for a project. When team members ask for a *comp* they usually mean that they would like a PDF or printout of the design.

Construct to Close: The second half of the rollout process, starting with content delivery and ending with cutover.

Construct to Close schedule: A project plan for the second half of the rollout process which is normally not finalized until the Plan and Define phase is complete.

Content: The assets used to produce a deliverable. (*See also* Assets.)

Content authors: Administrators of content management systems and digital asset managers who are responsible for building website structures and populating them with assets.

Content delivery network (CDN): This is a system of server nodes located across the globe, strategically placed to provide users with quick access and response time when searching for content.

CDN change: When the code that tells the CDN where to find content or where content will be saved within the CDN is updated to reflect a scheduled alternation.

Content lead: Experienced content authors who oversee a group of other authors as they work together on a project.

Content management system (CMS): A web application that allows nontechnical resources to build and manage websites using standardized templates and components.

Content optimization: Techniques the search team recommends to the creative team in order to increase the likelihood that their content will help the site to index high on search engines.

Content quality analysis: The process of reviewing the content of a digital project to make sure it matches the comps, wireframes, or copy docs as expected. This task is performed by the quality analysis team before clients or external parties test the project so that errors are identified and corrected in advance.

Content strategy: For the purposes of building a responsive website, this term is used to describe how the design team uses relative units, breakpoints, maximum widths, and other tactics to ensure that each page looks great regardless of viewport.

Content tracker: A spreadsheet that lists the specifications for every single asset expected to be delivered for a project, and is used to ensure that delivered assets match the specs. It also serves as a report so the team can review the status of every piece of required content.

Copy doc: The copy *document* is delivered from a copy writer and is a word processing file with the required text for the project. The text can include headlines, paragraph text, image captions, alt text, disclaimers, or anything else required to build the project.

Core team: Critical team members who need to actively and consistently engage in the tasks and communications of a project in order for it to succeed.

Critical path: Steps of the process that need to happen consecutively because the first task must be complete before the second can be completed, and so on.

Cutover: Also called *launch* and *go live*, this is the process of connecting a domain with an IP address and publishing it to the target audience.

Cutover activities: Tasks associated with the public release of a project that need to happen both prior to and on the day of cutover.

Cutover management: The act of managing the cutover activities.

Data architect: This is a type of developer who designs and manages the flow of data from end-to-end. If the project requires either a data source or data output, the architect will figure out how to pull the data in, filter it, merge it, present it, and/or push it out to a third party.

Defects: Also called *bugs*, defects are identified by checking the test deliverable against the original specifications. The delta between the two is an error (defect) that needs to be logged and corrected.

Deliverables schedule: A list of either actual deliverables or milestone activities.

Development environments: Levels of environments within the infrastructure in which project teams work to develop and test digital deliverables.

Digital asset management (DAM): A central repository of assets used to organize, manage, maintain, update, share, and distribute them to multiple projects.

Digital creative agency: Marketing and advertising agencies that specialize in the creative conception and graphic design of digital campaigns, but are not responsible for the assembly or development of the projects.

Digital production agency: Marketing and advertising agencies that specialize in the assembly and development of digital campaigns, but are not responsible for the creative conception or graphic design of the assets.

Disaster recovery: Policies, procedures, and systems put in place to recover vital technologies in the event of a natural or man-made disaster.

DNS record: The domain name system (DNS) is a record of domain names and their related IP addresses.

End-to-end testing: A phase of testing when the quality team ensures that the data sent on the front end is received as expected at the final destination.

Executive summary: A very concise summary of the project status prepared for management review.

Extended team: Stakeholders who are not be part of the *core* team, but have some level of interest in the outcome of the project.

Extensible markup language (xml): A computer language that can be read by web applications and is often used in place of a database solution, when one isn't available.

External site: This applies to links within a website that connect to points outside of the site.

Failover: Redundant hardware and servers that are in place and ready to take over in case the original systems fail.

Favicons: Standing for *favorite icons*, these are the tiny logos shown in browser tabs and in the bookmarks list of saved websites.

Final budget estimate: The final version of the budget that is presented to clients after all requirements have been captured.

Freak out: Something business owners do during a cutover, while the sites are in transition. This normally takes place after the domain mapping has been changed but before the 301s have been implemented.

Frontend programmer: A developer whose expertise is creating the code that supports the presentation layer of a project.

Full-service digital agency: A marketing and advertising company that can provide everything the client needs to produce a digital project. From concept and design to programming and implementation, the client need not hire any other partner to produce the desired outcome.

Functional requirements: Mainly refers to how the user interface will respond to the customer when they interact with the website, but also includes any requirements needed to fulfill the business needs.

Functional specifications: Comprehensive execution instructions and designs that will be used to implement the technical solutions.

Gap analysis: An assessment of the delta between the client's desired outcome and the current capabilities, conducted in an effort to identify project requirements.

Global footer: The area at the bottom of the page that stays in place and doesn't change no matter where the user is within the website.

Global header: The area at the top of the page that stays in place and doesn't change no matter where the user is within the website, which normally includes the navigation and various website identifiers such as the brand logo.

Global teams: When core team members are dispersed around the world and use web conferencing, phone calls, e-mails, and other solutions to communicate and manage a project.

Grid system: The number, size, and assortment of columns available to the designers.

Gutter: For digital projects, this term refers to the space to the left or right side of components, but it was originally used in print—referring to the space between two pages of a spread where the binding would join them together.

H1, H2, H3...: HTML tags that indicate the level of headings that are being displayed. H1 is the most important heading on the page, H2 the second most important, and so on. These tags are used by CSS developers to style the headlines, and are read by search engines.

Hamburger menu: This style of mobile site menu got its name because it actually looks like a tiny hamburger.

High-level budget estimate: A type of budget estimate provided at the beginning of a project, before the project manager has enough information to create a detailed budget—which makes it more of a guess.

Hot fix: When code is released into an environment outside of the normal release cycle.

Hover state: The graphic design change that takes place within text or images as users engage with them but don't actually click on them. (*See also* rollover state.)

Image naming convention: The actual names given to images when they are saved.

Image optimization: When art directors or asset managers create the highest resolution possible for an image confined to size and weight restrictions.

Image weight: Refers to the size of the image, often measured in bytes.

Information architect: This resource collaborates with graphics designers to develop the customer experience for digital projects conveyed through sitemaps and wireframes.

Infrastructure manager: The resource responsible for the overall stability of the platform, including environments, bill of materials, and regulatory processes.

Initial research: Investigation work conducted upon initial receipt of a new assignment in order to gather enough information to get the team started on first steps.

Initiation phase: This refers to the first six steps of the rollout process, from gap analysis to the statement of work tollgate.

Internal kick-off meeting: Once the scope of work and budget have been approved by the client, the project manager conducts this internal meeting where all assigned resources are briefed on the project and officially begin work.

Internal QA: Quality assurance checks that the internal test team performs before external parties begin their review, in order to identify and correct defects prior to outside exposure.

Issues log: This is a list of challenges that the core team isn't yet sure how to resolve or find a solution for.

IT bill of materials: A list of all the hardware, servers, memory, or other materials currently owned and operated by the host provider. This could include operating systems, network services, network hardware, rack hardware, or power and cooling supplies, among other things.

JPG: A compressed image file format with unlimited color options traditionally used for photographs.

Key performance indicators (KPIs): Measurable business objectives that prove how effectively a company's marketing and advertising solutions are working.

Kickoff: The same as *internal kick-off*, except it also includes all external team members and stakeholders.

Live area: The area visible on-screen without scrolling.

Localize: The practice of taking content from one market and altering it to suit another market's requirements. Language translations, marketing objectives, and technical capabilities are all part of the localization process.

Lower environments: All development environments except for production are considered *lower*.

Maintenance team: The team designated to maintain and update a website or other digital project on an on-going basis, for the duration of the life cycle of the project. (*See also* Sustain team.)

Masthead: The main image area at the very top of any page.

Maximum width: This may refer to either the largest size at which content strategists want images to be shown, or to column widths in terms of text area. These specifications are applied in order to manage different viewport layouts.

Meta data: Meta data is sometimes referred to as *data about data*, because it involves the organization of data. In terms of websites, meta data is included in the page source code and used by search engines to categorize and reference content.

Mobile detection: The process of identifying the type and size of device visiting a website (viewport) or other service so that the appropriate web experience can be published to the end user.

Mobile first: A methodology used in responsive design where the team designs for the mobile screen first and then expands on that experience for desktop users.

Natural search: Search results that are indexed through the efforts of the search engine's algorithms and web crawling process, and not through any advertising efforts.

Needs assessment: This part of the rollout process covers the gap analysis, workshop, and stakeholder list. It's through these critical early steps that we investigate and start to formally document project requirements.

Nonfunctional testing: A category of testing describing that which is performed on systems and processes, such as security, performance, disaster recovery, and failover.

Nontechnical wireframes: Wireframes that do not require functionality requirements or specifications because the pages they represent will only be used to publish static content and assets.

Online status reports: Project reporting tools that are hosted online and available to be updated by more than one team member, as no individual person owns the report.

Open source: Software that is available on the internet free of charge to the public.

Origin: A version of the production site that is used to authenticate published results because it pulls its content directly from the host server and is not sent through the content delivery network.

Package identification: Produced by the creative team, for the production team, so they know which assets will be grouped together for delivery.

Page load time: The elapsed time between when a user submits a call to a page and when the complete page is published in full.

Paid search: Search engine results based on keywords that advertisers have purchased, in order to guarantee prime indexing placement.

Performance testing: This nonfunctional test measures finished products against compliance standards for page load times, responsiveness, and system stress and spike stability.

Pixel: A single measurement point within a screen or graphic element.

Plan and Define: The first half of the rollout process which starts with the gap analysis and ends with the creation of the Construct to Close schedule.

Platform: The integrated software and hardware used to provide a digital solution.

PNG: A compressed image file format with limited color options traditionally used for line art, logos, and text-heavy images. PNGs also have a transparent background which makes them ideal for layering.

Preliminary budget estimate: A budget developed during the initiation phase of Plan and Define, once the business requirements have been based-lined, but before the planning activities have begun.

Pre-production environment: An area within the development infrastructure located just below the production environment, where final tests are conducted to ensure no known defects are released into production.

Production environment: The top level of the development infrastructure, where content and code are published to the target audience.

Programming language: This refers to the type of code developers use to create software—such as Java, PHP, C, C++, and C#.

Project scope: A detailed description of every task, resource, and deliverable that the client wishes to pay for.

Proportion-based grids: Used for responsive web design, this is a grid system that repositions itself based on viewport, rather than using fixed pixel-based dimensions.

Quality assurance (QA): The process the test team applies to ensure a project's outcome matches the original specifications.

Regression testing: Repeatable test scripts that are applied to development environments after new code is released into the system, to ensure that everything that used to work *still* works and was not adversely affected by the new code.

Relative units: A responsive design layout and measurement technique based on a ratio of viewport size.

Release: When a package of code is moved into a new environment.

Release management: The regulatory processes and systems that infrastructure managers use to ensure the stability of the development environments they oversee.

Release schedule: A schedule created and maintained by the release management team that communicates the dates on which new code will be accepted into each environment. Production teams work within these date parameters to create project plans and to schedule resources.

Request for proposal: An invitation for a potential supplier to provide the requesting company with a recommend solution and pricing for the desired outcome.

Request for quotation: An invitation for a potential supplier to provide the requesting company with pricing for a solution that has already been identified.

Responsive content: Assets for a digital project that either appear or disappear based on viewport.

Responsive design: A website that uses the same code to provide content to both desktop users and mobile users instead of building two different sites. For this approach a content strategy is created and utilized so that grids and assets adjust based on viewport.

Risk register: Refers to a list of possible issues that have been identified as having the potential to have a negative effect on a project. Those risks are then tracked and monitored so that the team is aware of the potential challenges and are therefore prepared to either avoid or mitigate them.

Rollout manager: A technical project manager who has a thorough understanding of the digital rollout process from gap analysis to cutover and can take responsibility to coordinate both the technical and business requirements of a project.

Rollover state: A change in the graphic representation of a page element (like a button or link) when it comes in contact with the user's mouse. (*See also* Hover state.)

RSS feed: The acronym actually stands for *rich site summary* but a more appropriate phrase that is often used is *really simple syndication*. It refers to a data push that allows publishers to automatically send updated data out on a regular basis so the receiving parties don't have to reach out to pull it in.

Sandbox activity: An environment for developers or authors to experiment with site production tasks in an effort to learn and investigate capabilities.

Scope freeze: This refers to the stage of the process when business and functional requirements are no longer being accepted into the current phase of the project.

Search engine optimization (SEO): Based on their knowledge of how content is crawled and indexed by search engines, teams that specialize in SEO make creative and structural recommendations to maximize a site's potential for a high index level.

Secondary navigation: While the global header navigation is static and appears at the top of each page of the website, *secondary* navigation only appears on certain pages within the global navigation that require additional guidance due to the number of available sub-pages.

Secure server: Servers that protect user's confidential information by encrypting the data captured when online forms are submitted.

Security testing: Quality assessment tests that are performed to ensure that secure servers are in place to protect confidential data exchange. They also ensure that the code used on a site has security measures in place to prevent unauthorized entities from getting access.

SEO agencies: Digital agencies that specialize in search engine optimization techniques and are hired to provide business teams with recommendations designed to increase their search result index levels.

Site speed: The amount of elapsed time it takes for pages to load and display all content and data.

Sitemap: A graphic outline of the website or application that indicates where each page will live in relation to the main navigation and within the site structure.

Sitemap footer: An outline of the website located within the footer navigation where each available page is represented by a hot link that users may click on to view the content.

Situation statement: A term that business analysts use to represent the *Needs Assessment* stage of the rollout process.

Social media strategy: The implementation plan developed and implemented by the social media team in order to achieve the client's social media business requirements.

Sprints: Groups of work packages that move through the design, development, and testing cycles together.

Stakeholder list: A comprehensive list of every core and extended team member associated with a project.

Statement of work: A document that provides a thorough explanation of the deliverables, resources, and costs that the client has agreed to.

Status report: A document that lists the key steps for a project, along with their associated activities and issues, so the team can comprehend where the project is in terms of the overall process and if there are any challenges blocking progress.

Styleguide: A document that specifies what the standard headlines, sub-headlines, paragraph text, forms, buttons, links, etc., all look like and what their transition behaviors are. It also specifies page background information, layer orders, and asset usage.

Sustain team: The team designated to maintain and update a website or other digital project on an on-going basis for the duration of the life cycle of the project. (*See also* Maintenance team.)

System integration testing (SIT): A phase of the overall test strategy that takes place after unit testing is complete, when all of the tools, components, and applications are tested together in an integrated environment. It is also the time that external software and database systems are aligned so that end-to-end testing can take place.

Technical project manager: A project manager who has a thorough understanding of the software development process, but who does not *necessarily* understand the activities, issues, and challenges that the business side faces.

Technical solution diagram: A graphic representation of the technical solution strategy.

Technical solution strategy: Comprehensive execution instructions and designs used to describe how the technical solution will be implemented.

Test cases: Very explicit directions written for the test team to ensure an accurate measure of approval is obtained, and as many defects as possible are identified.

Test strategy: A document that describes what the testing standards are, what *will* and will *not* be included in the test cycle, the criteria that will be used to determine if the test cycle has been successfully completed, the method for reporting defects, and the proposed schedule.

Testing environment: An environment within the infrastructure that is designated for testing procedures only—no programming activities.

Ticketing system: Software built specifically to open and track defects for digital projects.

Time to Live (TTL): The amount of elapsed time from when a website administrator releases new content or code to the time it's published around the world.

Tollgate: A decision point within a process when the business owner decides whether or not to proceed with the project as defined by its current scope.

Transition to operations: The last official step in the rollout process when the team that built the project trains and works alongside the operations team (sustain team) to provide support before the build team moves on to another project.

Unique visits: The number of individuals who visit a website during a period of time, regardless of how often they visit.

Unit testing: Testing conducted prior to SIT on individual requirements, tools, or applications associated with a project.

URL strategy: The techniques, trends, guidelines, and recommendations that the SEO team implements to ensure that URLs are written to take advantage of the latest ranking factors.

Use cases: A requirement-defining methodology for which statements are written in first-person in order to get a user's perspective on what they might want the deliverable to do for them. For example, "I need the store locator tool to identify and list the stores closest to my current location, using my mobile device's GPS."

User acceptance testing (UAT): The final phase of testing when the business owners thoroughly review all content and development deliverables before giving their approval to launch.

User actions: Ways in which the target audience interacts with the digital deliverable, including swiping, clicking, reading, watching, hovering, etc.

Vector images: These types of graphic assets are ideal for responsive web design because they're able to maintain the same level of quality regardless of size.

Version control: The process of updating the naming convention of a document that has already been distributed, so that it's clear that the viewer is looking at the latest update.

Viewport: The actual screen for digital devices such as laptops, mobile phones, or tablets.

Virtual host: When multiple domain names are contained and published from a single server source, allowing the group to share hardware, software, and processes.

Waterfall approach: A development methodology where all work packages move through the design, development, and testing cycles together without allowing any one package to move ahead of the others.

Web application: A software program that can run in a browser without having to be downloaded onto a device.

Wireframes: Line drawings that show the various elements of a digital project, typically used for planning a site's structure and functionality.

INDEX

Note: Page numbers followed by "*f*" indicates a figure and "*t*" indicates a table.

security, 155–156
SIT, 154, 191–198
strategy, 150–156, 151*f*
UAT, 154, 198–202
unit, 153–154
wrap-up presentation, 170
tickets/ticketing system
  assignee, 193
  attachments, 193–194
  cutover, 217–218
  dashboard review, 199–201
    critical tickets, 200–201
    issue type, 201
    tool-specific tickets, 201
  described, 191, 192*f*
  description, 193
  issue type, 192–193
  labels, 194
  priority, 193
  project, 192
  summary, 193
time zones, 81
translations, global rollouts, 206
Trip Planner tool, 142–144, 162
  development schedule with,
    142, 143*t*
TTL (time to live), 217

unit testing, 153–154
URLs
  blank fields, 214
  error messages, 220–221
  external links, 213
  removing duplicate, 212–213
  search engine optimization
    (SEO), 119, 120*f*
  special characters on, 213
  301 redirects, 210–214, 216
user acceptance testing (UAT), 154,
    198–202
  approach to, 199
  dashboard review, 199–201, 200*f*
  final review and schedule, 162,
    163, 164–165, 165*t*

guidelines, 201–202
meeting schedule, 201
prep meeting agenda, 198, 198*f*
roles and responsibilities, 199,
  200*f*
schedule, 198–199

validations, 147–148
vector images, 107
verbal approval, 58
version control, 79–80
vertical status reports, 75, 76–77,
  77*f*
video, alt tags, 118–119
viewport, 106
visitors, 114–115

web developer, 3
web programmers, 3
WebEx, 81–82
website
  responsive design, 5–6
wireframes
  described, 91–92
  footer, global navigation, 98–99
  functional specifications *vs.*, 91
  header, global navigation, 92–98
  homepage, 99–102, 100*f*, 101*f*
  need of, 92
  schedule for, 60, 61–66, 67
  wrap-up presentation, 169
working remotely, 171
workshop
  agenda, 25, 26–27
  conducting, 25–27
  invitees, 25–26
  purpose of, 25
  scheduling, 25
wrap-up presentation, 166–170

xml. *See* extensible markup
  language (xml)